HARVESTING RAINWATER FOR YOUR HOMESTEAD

Transform Your Home and Farm with Rainwater to Uncover Self-Sufficiency Secrets and Adapt Your Ideal Tank Without Calculations Ensuring Clean Unchlorinated Water in Every Setting from Urban to Rural Budget-Friendly

PATRICK P. ELIASON

✦ HERE IS YOU FREE GIFT!
⬇ SCAN HERE TO DOWNLOAD IT

1. Discover in "**My Solar Powered Lifestyle**" how to drastically reduce your bills and live in harmony with nature.
2. With "**EasyRain for Revolutionary and Sustainable Homesteading**", learn innovative techniques to make the most of every drop of rainwater.
3. Combine the power of the sun and water for a completely self-sufficient and eco-friendly life.

⬇ SCAN HERE TO DOWNLOAD IT FOR FREE

© Copyright 2024 - All rights reserved.

The content contained within this book may not be reproduced, duplicated or transmitted without direct written permission from the author or the publisher.

Under no circumstances will any blame or legal responsibility be held against the publisher, or author, for any damages, reparation, or monetary loss due to the information contained within this book. Either directly or indirectly.

Legal Notice:

This book is copyright protected. This book is only for personal use. You cannot amend, distribute, sell, use, quote or paraphrase any part, or the content within this book, without the consent of the author or publisher.

Disclaimer Notice:

Please note the information contained within this document is for educational and entertainment purposes only. All effort has been executed to present accurate, up to date, and reliable, complete information. No warranties of any kind are declared or implied. Readers acknowledge that the author is not engaging in the rendering of legal, financial, medical or professional advice. The content within this book has been derived from various sources. Please consult a licensed professional before attempting any techniques outlined in this book.

By reading this document, the reader agrees that under no circumstances is the author responsible for any losses, direct or indirect, which are incurred as a result of the use of information contained within this document, including, but not limited to, — errors, omissions, or inaccuracies.

TABLE OF CONTENTS

INTRODUCTION 6

CHAPTER 1
INTRODUCTION TO RAINWATER HARVESTING 8
Origins and Evolution of Rainwater Harvesting 8
Why Harvest Rainwater 11
Sustainability and Conservation Aspects 12

CHAPTER 2
UNDERSTANDING THE BASICS 15
The Water Cycle and Rain Formation 15
Quality of Rainwater 17
Factors Influencing Rainwater Harvesting 20

CHAPTER 3
LEGAL CONSIDERATIONS 23
Overview of Regulations Worldwide 23
Seeking Permissions and Adhering to Local Laws 27

CHAPTER 4
ASSESSING YOUR RAINWATER POTENTIAL 30
Calculating Roof Surface Area 30
Understanding Rainfall Patterns and Calculations 32
Potential Storage Needs 34

CHAPTER 5
COMPONENTS OF A RAINWATER HARVESTING SYSTEM 36
Catchment Surface: Roofs and Beyond 36
Gutters and Downspouts 37
First Flush Systems 38
Storage Tanks and Barrels 38
Filters and Screens 39
Overflows and Outlets 40

CHAPTER 6
RAINWATER QUALITY AND PURIFICATION 42
Common Contaminants 42
Filtration Methods 44
Purification Techniques: Boiling, UV, and Chlorination 46
Testing Water Quality 47

CHAPTER 7
DESIGNING YOUR RAINWATER HARVESTING SYSTEM 49
Basics of System Design 49

Gravity vs. Pump Systems...51
Above-Ground vs. Underground Storage..52
Scalability and Future-Proofing Your System 53

CHAPTER 8
STEP-BY-STEP PROJECTS FOR THE HOMESTEADER..55
Simple Rain Barrel Collection: A Deep Dive into the Basics..55
Advanced Rainwater Collection with First Flush..56
Underground Rainwater Storage System: Harnessing Earth's Bounty.....................................58
Integrating Rainwater into Home Plumbing: A Comprehensive Guide....................................60
Creating a Rain Garden: A Flourishing Oasis of Sustainability...63
Rooftop Garden Integration: A Verdant Oasis Above..65

CHAPTER 9
USING COLLECTED RAINWATER EFFECTIVELY...68
Garden and Agricultural Irrigation 68
Indoor Uses: Toilets and Laundry 70
Drinking and Cooking: Ensuring Portability 71
Emergency Preparedness..73

CHAPTER 10
MAINTENANCE, TROUBLESHOOTING, AND COMMON CHALLENGES 75
Regular Maintenance Tasks 75
Addressing Common Issues: Algae, Mosquitoes, Leaks 77
Winter Care and Freeze Protection 79
Expanding or Upgrading Your System 80

CHAPTER 11
SUSTAINABILITY AND ENVIRONMENTAL BENEFITS..82
The Role of Rainwater Harvesting in Conservation 82
Combining with Other Sustainable Practices: Greywater, Composting, etc............................85
Impacts on Local Water Systems and Aquifers 89

CHAPTER 12
COMMUNITY AND EDUCATIONAL OUTREACH..94
Building a Community Harvesting Project 94
Educating Neighbors and Local Schools 97

CHAPTER 13
BEYOND THE BASICS..99
Advanced Filtration Systems..99
Smart Monitoring with Digital Tools...100
Incorporating Solar Pumps 101
Landscape Design for Optimal Rainwater Use...102

CONCLUSION..105

INTRODUCTION

A single raindrop—seemingly insignificant on its own—makes a mighty impact when combined with billions of others. As these clear pearls fall from the sky during a refreshing spring shower, they unleash a cascade of abundance. Trees unfurl their leaves to soak up the moisture while seeds below the soil begin to sprout. The parched earth drinks its full, and rivers swell with silt-rich runoff. Wildlife flock to quench their thirst, and balance is restored in nature. Such is the beauty of the water cycle, transforming vapor into precipitation that sustains all life.

Yet, in the backdrop lurks a paradox. While essential for existence, freshwater faces mounting scarcity around the globe. Many of our miles-deep aquifers are drained faster than they are replenished. Pollution plagues our waterways as demand rapidly outpaces supply. Countries quarrel over access to rivers and lakes that cross borders, while corporations privatize this once-common resource for profit. The victims are ultimately everyday people and delicate ecosystems left high and dry.

It seems we have taken water for granted, putting ourselves on a collision course with crisis. But what if the solution was right above our heads? An ancient yet cutting-edge innovation beckons us—harvesting the rain. From the erecting of primitive clay pots by farmers in India over 4,000 years ago to the advanced rainwater collection systems at modern green buildings, humanity has tapped into this sustainable source for ages. While more ubiquitous in past eras, rainwater harvesting is making a comeback in the face of adversity. And homesteaders, ever the pioneers, are leading the charge once more.

As our lifestyles grow exceedingly disconnected from nature, homesteading calls us back to our roots. It embodies self-sufficiency and independence; harmony between ecology and economy. At its core, homesteading lets us fashion the world we wish to inhabit—one of health, balance, and beauty using what the land freely offers.

Rainwater harvesting then becomes the cornerstone for homesteads to thrive while trending towards sustainability. It puts us in sync with nature's hydrologic rhythms instead of dominating them. It teaches patience and preparedness, ingenuity, and vigilance. And it distills life's essential ingredients down to a basic ritual of gathering what falls from the firmament. The principles spill into all facets of homesteading—growing food, raising animals, compact living, and disaster readiness.

While a global crisis looms, homesteaders, new and veteran, have cause for inspiration. Our ancestors prevailed in harsher climes, and so can we, integrating the best of simplicity from bygone eras with modern innovations. Harvesting Rainwater for Your Homestead pools from this wellspring of wisdom, funneling it into a guide that is both pragmatic and empowering. It ushers even novice homesteaders to embrace rainwater harvesting as a fulfilling lifestyle and ethic.

- Through the course of this book, we traverse such topics as:
- The storied history of rainwater harvesting around the world
- Core concepts every rainwater harvesting enthusiast must grasp
- Legal considerations based on where you call home
- Step-by-step guidance for installing complete systems with illustrations
- Purifying and maintaining high water quality
- Troubleshooting common problems
- Ideas for using rainwater across your entire homestead
- Future-proofing through modular design

While centered on homesteaders, this book proves useful for a spectrum of readers. Urban farmers can adapt the principles to green rooftop gardens. Suburban families may install rain barrels to bypass municipal water for their lawns. Communities can rally together to build large-scale rain catchment and storage to combat drought. And anyone with a reverence for water can discover how to respectfully harness nature's gift.

Ultimately, Harvesting Rainwater for Your Homestead serves as a practical crash-course and timeless reference guide. It awakens us to water scarcity while offering actionable solutions. The journey promises to transform how we perceive, preserve, and partake of this precious resource. Our homesteads become the fertile training ground where a thriving future takes root with every drop we rescue from runoff and every system we engineer. And from here ripples a message the world desperately needs—that sustainability is possible one raindrop at a time.

CHAPTER 1
INTRODUCTION TO RAINWATER HARVESTING

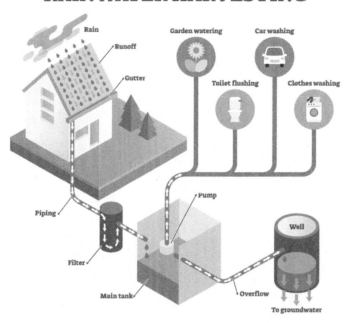

Rainwater harvesting refers to the ancient yet innovative practice of collecting and storing rainwater for various uses. With roots tracing back thousands of years, rainwater harvesting continues to be highly relevant, especially for sustainable living. This introductory chapter provides a historical overview of rainwater harvesting, highlights its myriad benefits, and emphasizes its environmental and sustainable aspects. It sets the overall context for the book.

Origins and Evolution of Rainwater Harvesting

The practice of collecting and storing rainfall has ancient roots across the world's civilizations. Societies devised ingenious rainwater harvesting systems aligned with regional climates and water needs. Over centuries, methods were refined and spread to make rainwater utilization ubiquitous globally.

Evidence of Rainwater Harvesting in Ancient Civilizations

The ancient Mediterranean civilizations demonstrated advanced rainwater harvesting skills. Archaeological evidence indicates rainwater usage in the Minoan civilization on Crete from 2000 BCE onwards. Elaborate drainage systems transported water. The ancient Romans built upon this expertise, engineering complex aqueducts and cisterns to supply their cities. Roman cisterns were typically brick-lined underground structures filled by runoff from roofs and paved surfaces. The ancient Persians, too, relied on an intricate underground drainage network called the qanat system that tapped into rainwater.

Similarly, the ancient Indian civilization valued water greatly. Archaeological and scriptural evidence shows sophisticated water harvesting traditions tracing back over 9000 years ago to the Neolithic Indus Valley Civilization. Large public baths and advanced drainage systems indicated rainwater harvesting knowledge. Ancient Indian treatises prescribed detailed guidelines for water management. Emblematic step-wells were built to access seasonal fluctuation in groundwater recharged by rainwater. Water harvesting was essential for Indian agriculture to flourish.

Spread of Rainwater Harvesting Knowledge Through Trade and Conquest

The extensive trading networks across Asia and the Middle East enabled the sharing of water harvesting techniques. As travelers journeyed along the Silk Route from 200 BCE to 200 CE, they exchanged not just goods but ideas. Chinese records from the Western Han dynasty mention rainwater harvesting. The technique likely spread east from Persia. The arid Middle East also relied on traditional water harvesting methods. Urban dwellers in Yemen harvested runoff in cisterns for daily use and irrigation. Simple brush dams trapped moisture in Jordan. Such examples showcase rainwater harvesting's spread and adaptation to new regions through trade and contact.

Infiltration and irrigation methods also spread through conquest. The Greek and Roman conquests spread irrigation and drainage methods within their empires. As Islamic empires rose from the 7th century CE onwards, they carried water management ideas from the Middle East westward into North Africa and Spain. Mediterranean rainwater harvesting traveled to the Americas with the Spanish colonists in the 16th century CE.

Evolution of Rainwater Harvesting in Asia

In China, rainwater harvesting evolved with the introduction of Mediterranean methods like grids by Jesuit missionaries in the 16th century. The local people merged imported and native techniques for optimum results. The Vietnamese adapted techniques from China starting in the 15th century CE, creating systems like communal village tanks. The jhiri, unique to western India, emerged around the 17th century CE. These were cone-shaped rainwater storage pits that funneled runoff via channels into an underground tank.

Rainwater Harvesting Arrives in the Americas

As European colonists arrived in the Americas from the 16th century CE onwards, they brought rainwater knowledge from back home. As early as 1515 CE, farmers in what became the US Virgin Islands constructed stone-lined cisterns to store rainwater. These large underground tanks collected runoff from roofs. On the mainland, early English settlers in Massachusetts relied on rainwater harvesting, given the lack of freshwater springs. Methods were rudimentary initially, consisting of storing water in barrels.

In the Caribbean, slaves on the plantations became adept at rainwater harvesting. They maintained cisterns and improved collection with methods like mortar and plastering to prevent leakage. The locals continued these practices after Emancipation. Rainwater harvesting became the predominant source for islanders, given the scarcity of freshwater and lack of piped water. Over time, underground concrete cisterns substituted old stone ones as the technique evolved locally.

Perfection of Rainwater Harvesting in Australia

British settlers experimented with rainwater harvesting to contend with Australia's dry climate from the late 18th century CE onwards. But the systems were basic, with leaking wooden tanks. In the 1830s, Australians started constructing more durable brick-lined tanks set into the ground to prevent evaporation. These tanks had a concave, dome-shaped roof with an opening for rainfall entry and an access hole above for drawing water. This basic Australian 'egg-shaped' underground tank design proved efficient, easy to build from available materials, and durable. It was widely adopted across rural Australia by the 1860s. Families also invested in large, elaborate above-ground tanks. Rainwater became the predominant water source in Australia by 1900 CE.

Gradual Refinement of Collection Techniques

Early rainwater harvesting systems were prone to contamination since storage tanks were open. Around the mid-19th century, filters were introduced to improve water quality. Basic rope and wire mesh filters fitted to inflow pipes minimized debris and insects. More sophisticated methods like first-flush diversion were invented. These bypassed the initial dirtier portion of rainfall. Settling tanks and slow sand filtration further improved stored rainwater quality. Chlorination and boiling were also adopted. Such advances made rainwater potable. The Australian government's promotion of rooftop rainwater tanks from the 1990s onwards helped improve urban water security further.

Evolution of Storage Methods

Pre-modern rainwater harvesting relied on underground brick, stone, and concrete cisterns or large wooden surface tanks. Between the World Wars, bolted steel and pressed steel tanks gained popularity, given Prefabrication enabled mass production. Plastics like polyethylene were adapted for rain storage from the 1950s onwards. These flexi-tanks were portable and affordable. Ferrocement tanks were introduced in Thailand and spread across Asia and Africa. The material was flexible, durable, and inexpensive. Such modern storage options made rainwater harvesting more feasible and economical.

Adoption of New Collection Elements

Simple pipes historically sufficed to transport rain from surfaces to storage. However, specialized fixtures improved collection efficiency. Rain chains, invented in 18th century Japan, broke descending water into individual drops to aerate and decelerate it. First patented in 1896, roof gutters enabled collection from larger roof areas. Downspouts carried rain to storage neatly. Parapet collection systems caught rainwater overflowing from rooftops for additional yield. Such enhancements optimized rainwater harvesting.

Revival and Global Spread of Rainwater Harvesting

By the mid-20th century, piped municipal water had reduced urban rainwater harvesting in many parts of the world, except Australia. However, growing water scarcity revived interest in traditional methods. Owing to its decentralized nature, rainwater harvesting provided water security resilience. Around the 1990s, Germany encouraged rainwater harvesting with subsidies. The United Kingdom made it compulsory for new buildings. Singapore, Japan, and China incentivized rooftop harvesting in cities. Rainwater harvesting rebounded as a contemporary urban water strategy.

Harvesting Rainwater in Modern Times

Today, rainwater harvesting has been embraced worldwide, from private homes to corporate campuses. Architectural designs seamlessly incorporate rain collection elements. Cities promote small-scale neighborhood systems to reduce stormwater outflow. Developing communities use simple rainwater techniques to improve health and water security. Rainwater harvesting is an integral part of the modern green infrastructure. Contemporary methods build upon millennia-old water wisdom to innovate sustainable techniques adapted to 21st-century climates and lifestyles.

Why Harvest Rainwater

Rainwater harvesting offers far-reaching benefits that make adopting the practice worthwhile. From increasing water security to reducing bills, the incentives span from the personal to the collective realms. This section details the diverse reasons motivating individuals and communities worldwide to turn to harvesting rainwater.

Rainwater as a Pristine Freshwater Source

As rain falls through clean air, it remains free of minerals, chemicals, and contaminants that often impair ground and surface supplies. Rainwater has a neutral pH, soft texture, and no smell. Hence, it is inherently pure and ideal for drinking, cooking, bathing, cleaning, gardening, livestock rearing, and many other domestic and agricultural uses. Harvesting rainwater provides a renewable source of high-quality water right where it is needed.

Supplementing Municipal Water Supply

Municipal water requires extensive transportation, treatment, and infrastructure supplied by utilities. Particularly in urban areas, supplying piped water entails high energy expenditure. Harvesting rain reduces the demand for municipal water by meeting needs like gardening, washing, toilet flushing, and outdoor uses. This conserves and takes the pressure off tap water, enabling a dual water supply system. Depending on rainfall, between 20 to 50 percent of household water can be met with rainwater.

Mitigating Water Scarcity

Increasing episodes of drought make stored rainwater a buffer against uncertainty in many water-scarce regions. Rainwater systems make communities resilient by ensuring water availability even when regular supplies are disrupted. From Australia to the Virgin Islands, rainwater harvesting provides a decentralized water source independent of changing climate conditions. This self-reliance also reduces inter-community conflicts over water.

Enhancing Water Security in Rural Areas

Beyond urban advantages, rainwater harvesting offers great benefits for rural homesteads lacking piped water access. Stored rainwater offers a convenient, close water source compared to distant surface or groundwater. Women and children, in particular, avoid trekking daily for water. Rainwater quality is also safer than untreated surface water, which is prone to contamination. Harvesting rain dramatically improves water security and quality of life.

Providing Emergency Water Supplies

Stored rainwater serves as a reliable backup source in emergencies when piped water is disrupted, like during natural disasters or accidents. Fire departments recommend keeping rainwater reserves to quickly douse fires, especially in rural locations. The ability to use rainwater for all purposes makes it an ideal emergency supply.

Reducing Runoff, Erosion and Flooding

Untapped rainwater generally flows across paved surfaces as runoff. But rainwater harvesting stems from this wasteful outflow. Retaining water in distributed storage reduces runoff volume and velocity. This prevents soil erosion and flooding downstream during heavy rainfall. Stored water gradually recharges underground aquifers, too. Water is retained closer to where it falls.

Promoting Self-Sufficiency

Harvesting rainwater promotes self-reliance by minimizing dependence on external utilities or government water sources. This ethos of independence resonates with homesteaders seeking a self-sufficient lifestyle. Installing rainwater systems enables greater control over personal water security. The satisfaction of meeting needs sustainably through rainwater also has emotional value.

Lowering Household Water Bills

Using stored rainwater for all non-potable uses substantially reduces tap water consumption. For the volume displaced, utility bills fall noticeably every billing cycle. The City of Tucson, Arizona, estimates that households relying on rainwater for landscape irrigation alone can save $400 annually. Rainwater harvesting yields ongoing cost savings.

Improving Yield from Gardens and Small Farms

The natural softness and gentle pH of rainwater make it ideal for watering gardens, plants, and crops. Denatured salts, chlorine, and fluoride in tap water can restrict soil nutrient absorption and damage plants when used regularly. Rainwater eliminates such risks, boosting gardens and farms' health and yield over the long term.

Easy Maintenance Requirements

Basic debris screens on inflow pipes keep rainwater clean. Beyond periodically inspecting and cleaning storage tanks, minimal maintenance is required. Harvesting systems are simple and passive, without energy requirements or complex mechanisms. Maintenance needs are easily manageable.

Suitable Across Climate Zones

Myths persist that rainwater cannot be harvested in cooler or arid regions. However, ample rainwater can be collected across climate zones to significantly offset external water needs. For example, Tucson, Arizona, receives 12 inches of rain annually, sufficient for over 50 percent of non-potable uses. Similarly, UK houses can source one-third of their needs from rainwater.

Sustainability and Conservation Aspects

Apart from the direct household benefits, rainwater harvesting also generates broader environmental and social gains that make it integral to sustainability. By adopting a holistic water cycle approach, rainwater harvesting provides ecological solutions while fostering community self-reliance.

Mitigating Water Scarcity

As climate change disrupts weather patterns, rainwater harvesting buffers against uncertainty by decentralizing urban water supply. Large cities can reduce their burden on rural watersheds. Melbourne met 17% of its needs through rainwater during its Millennial drought. Rainwater harvesting takes the pressure off the contested surface and groundwater reserves.

Storing rainwater makes communities resilient against droughts by ensuring supply even if regular sources are disrupted or rationed. Harvesting rain reduces dependence on rural bodies of water that may become stressed due to urban demand. Urban water self-reliance is an important sustainability strategy.

Improving Water Conservation

By tapping rainfall as a diffuse resource, dependence on centralized infrastructure and utilities is reduced. Using rainwater lessens the usage and contamination of freshwater sources. Designing urban landscapes to maximize rainwater catchment promotes a conservation culture. Residents become conscious of water cycles.

Seeing rain barrels and cisterns provides a tangible reminder of water as a precious resource. It reinforces saving water and averting wasteful usage. Rainwater harvesting makes water conservation a way of life.

Stormwater and Flood Management

Impervious surfaces like roads and parking lots generate large stormwater volumes. Harvesting rain in distributed storage reduces urban runoff velocity and volume. This controls downstream erosion and flooding while also recharging groundwater. Rainwater systems thus mitigate infrastructure costs of stormwater management.

Stored rainwater in the soil allows gradual percolation into aquifers, elevating water tables. Reduced surface runoff prevents soil loss and damage to streams from heavy water surges during storms. Rainwater harvesting aligns urban design with natural hydrology.

Enhancing Local Climate and Environment

Rainwater harvested in soils and open water bodies moderate ambient temperatures through evaporative cooling. Gardens nourished by rainwater create microclimates through transpiration. Rainwater thus provides passive environmental benefits that improve urban habitats and reduce the heat island effect.

The cooling effect of water-rich soil and vegetation improves thermal comfort and air quality in built-up areas. Site-based rainwater harvesting adds much-needed pockets of green infrastructure in cities and towns to counter the effects of extensive paving.

Energy Conservation

Large-scale water supply entails huge energy usage for pumping, transport, filtration, and distribution. Relying more on rainwater reduces this energy expenditure, given its passive collection. Energy policy increasingly prioritizes such appropriate technology like rainwater harvesting.

As pumping and moving water over long distances requires tremendous energy inputs, localized rainwater systems conserve electricity and fuel. This makes water and energy usage align more closely with ecological cycles.

Sustainable Urban Development

Rainwater harvesting represents an eco-friendly, deurbanized approach to basic needs provision, in contrast to resource-intensive centralized systems. Mainstreaming rainwater harvesting facilitates the sustainable design of neighborhoods and cities, improving ecological resilience.

Making rainwater harvesting mandatory for new construction steers development onto a low-impact path. The sustainable water ethos then gets built into urban form and lifestyle itself. Rainwater harvesting enables urbanization to co-exist with water conservation.

Self-Sufficiency and Community Resilience

By making communities less dependent on external utilities for water, rainwater harvesting promotes self-reliance, which is especially crucial in times of natural disaster. Neighborhood networks allow the sharing of knowledge and resources. Harvesting rainwater enables the building of disaster-resilient communities.

Community-based rainwater harvesting makes neighborhoods more self-contained for basic water needs rather than sourcing from distant reservoirs. This spirit of self-help and cooperation builds social capacity to withstand emergencies better.

Empowering Women

Rainwater systems reduce the burden of water collection, which is often imposed on women and children. Girls' education and family health improve when the time spent fetching water reduces. Access to water also empowers women economically through homestead enterprises.

Freeing up time spent carrying water allows women to invest in education, livelihoods, and childcare. Reduced drudgery provides gender equity benefits. Using rainwater raises women's status as water managers and breadwinners.

Water Quality

Rainwater eliminates risks of contamination present in surface and groundwater, which require extensive treatment. Pathogens or salts impairing health and agriculture are avoided. The soft, fresh rainwater also enhances skin and hair health compared to hard tap water.

With virtually no dissolved solids, the clean rainwater prevents scale build-up in pipes and staining. Bathing in rainwater improves hair smoothness and skin conditions compared to treated municipal supplies. The health benefits are noticeable.

Low Environmental Footprint

Rainwater harvesting has minimal ecological impact compared to massive centralized water systems requiring intensive resource and energy inputs. Small-scale rainwater harvesting provides a low-carbon means of meeting basic water needs while rejuvenating the environment.

The embedded energy in rainwater systems is negligible compared to large dams, pipelines, and treatment plants. Rainwater harvesting allows growing populations to increase water availability without negatively impacting nature.

CHAPTER 2
UNDERSTANDING THE BASICS

Water sustains all life. As raindrops fall from the sky, cascading down gutters and flowing into collection tanks, this elemental elixir nurtures gardens, quenches animals, and hydrates homes. Tracing the origins of rainwater connects us to the profound hydrologic cycle that circulates water continuously between land, sea, and sky. Understanding the journey of a raindrop gives insight into rainwater's unique properties as a pure, soft, mineral-rich resource. Beyond appreciating rain's quality, successful rainwater harvesting depends on carefully considering factors like local rainfall patterns, collection area size, storage capacity, and usage needs. By integrating knowledge of water's path from sky to soil and judiciously applying site-specific planning, rainwater harvesting allows us to consciously capture and utilize this renewable gift. This chapter explores the fundamentals underpinning effective rainwater collection for homesteads and households seeking self-sufficiency.

The Water Cycle and Rain Formation

The hydrologic cycle is the continuous circulation of water between the earth's surface, atmosphere, and back again. This delicate global dance of evaporation, condensation, and precipitation connects all life. Gaining awareness of these processes illustrates how rain forms and falls, providing essential context for rainwater harvesting practices.

The Sun's Radiance Powers Evaporation

The sun provides the energy that drives evaporation, transforming liquid and frozen water on the earth's surface into invisible vapor. This upward transfer of water can occur from oceans, lakes, rivers, moist soil, and transpiration from plants. Vegetation releases significant moisture into the air through tiny pores on leaves called stomata. This gaseous water remains suspended in the atmosphere until conditions cause it to condense back into liquid form.

Condensation Forms Clouds

As moist air rises and subsequently cools, water vapor condenses onto tiny airborne particulates. These suspended particles, such as dust, soot, and salt crystals, act as condensation nuclei, allowing vapor to transform into cloud droplets. As more moisture condenses, these tiny cloud droplets accumulate, gradually forming visible puffy cumulus and layered stratus clouds. The differing shapes and textures of clouds signal changing weather patterns to the observant eye.

Saturation Leads to Precipitation

Within a cloud, when the millions of tiny droplets become too heavy, they combine into larger drops. As more moisture condenses onto these drops, they become heavy enough to overcome the upward drafts that previously suspended them. At this point of saturation, the drops fall as precipitation. Variations in temperature dictate whether rain, snow, sleet, or hail occurs as the drops continue their downward journey.

Elevation Changes Impact Raindrops

During the descent, conditions like temperature and humidity levels will change as elevation decreases. This can potentially cause some raindrops to partially evaporate, cycling moisture back into vapor form if conditions along the path become warm and dry. However, eventually, most drops reach the earth's surface, replenishing the moisture in lakes, rivers, aquifers, soil, and plants to continue the hydrologic cycle.

The Cycle Continuously Recirculates Water

With this balanced circulation, water gets transferred perpetually around the planet in a closed-loop system. The processes of evaporation, condensation, and precipitation maintain life by redistributing freshwater stores. Gaining awareness of nature's hydrologic cycle provides meaningful context for rainwater harvesting practices.

Harnessing the Cycle Through Collection

Humans interrupt the balanced natural cycle by consciously capturing and storing rainwater for beneficial use. However, with mindful collection and efficient utilization of harvested rain, equilibrium can be sustained even while rerouting some water for human purposes.

Radiant Energy from the Sun

The sun provides the crucial radiant energy that powers evaporation and transpiration, beginning the upward transfer of water into the atmosphere. Solar radiation heats the earth's surfaces, including oceans, seas, lakes, rivers, moist soil, and vegetation. This input of thermal energy causes liquid water to transform into invisible water vapor in a gaseous state through the process of evaporation.

Plants also lose substantial moisture from leaves and stems through tiny pores called stomata. This evaporation of water through plant transpiration is an important contributor to atmospheric moisture. Without the sun constantly supplying warmth, this movement of water into the air would cease. Solar radiance provides the essential heat that initiates the hydrologic cycle.

Global Conveyor Belt

The continuous circulation of water around the planet functions akin to a complex global conveyor belt. Heated near the equator, vapor rises, condensation occurs, and precipitation delivers downslope flows toward the poles. Ascending air currents transport this moisture while descending cool and dry air masses in temperate zones spread equatorial waters poleward through rain, rivers, and ocean currents.

This interconnected system of circulating freshwater stores maintains habitable conditions across the earth's varying climates. Gaining awareness of our planet's hydrologic 'conveyor belt' provides a meaningful context for rainwater harvesting as both a local and global practice.

Phase Change Transforms Water

Water undergoes important phase changes within the hydrologic cycle, shifting between solid, liquid, and gas states. The sun melts ice into liquid water and provides the energy for evaporation or sublimation into gaseous water vapor. Rising cooled air causes condensation back into liquid droplets that can freeze into ice crystals depending on conditions in the atmosphere.

Temperature and pressure prompt water's phase transformations as it circulates between the earth's surface and the sky. Comprehending these states of matter helps grasp how invisible vapor materializes as clouds and precipitation.

Biotic and Abiotic Factors Interact

Complex interactions between biotic living components like vegetation and animals and abiotic non-living elements like soil, water, and air drive the hydrologic cycle. Transpiration from plants, evaporation from bodies of water, solar energy, gravity, and air temperature all play roles.

This dynamic system strives for equilibrium even as human activities cause variations. Appreciating these numerous influencing factors provides insight into water's journey.

Ascending and Descending Limbs

Meteorologists studying the hydrologic cycle often describe 'ascending' and 'descending' limbs to model atmospheric circulation. Moisture enters the rising limb through evaporation and transpiration, moving as vapor in air currents aloft. Upper atmosphere cooling results in condensation into clouds.

Precipitation forms the descending limb, returning water downslope through rain, snow, and other forms. At the surface, it may enter lakes, rivers, or aquifers to repeat the cycle. Observing this general upward and downward movement reveals the cycling process.

Quality of Rainwater

The purity and softness of fresh rainwater augment its value for drinking, cleaning, gardening, animal care, and numerous other household and smallholding applications. As precipitation falls through the atmosphere, it remains free of most impurities that compromise ground and surface supplies. The low dissolved mineral content reduces scaling and extends the soap's effectiveness. However, airborne pollutants, contaminated catchment areas, storage methods, and other factors can affect quality. Careful collection, diversion of initial contaminants, filtration where required, and optimal storage preserve rainwater's beneficial attributes. Evaluating both the potential risks and the best practices to obtain high-quality water informs appropriate usage.

Natural Softness

A primary advantage of rainwater is its natural softness and low dissolved mineral levels. Water hardness relates to the presence of calcium and magnesium ions. Groundwater picks up these alkaline components, filtering through layers of limestone, chalk, and other sedimentary deposits. Municipal water gets hardness from both the source and chemical treatments during processing.

The low ionic content of rainwater prevents the buildup of scale deposits in pipes, tanks, and appliances. This saves energy and maintenance costs over time. The soft quality also prevents mineral residue on surfaces, increasing cleaning effectiveness. With less insoluble compounds, rainwater improves lathering and rinsing with soaps and detergents.

Purity from the Sky

Raindrops form around minute aerosol particles in the relatively clean moisture of atmospheric clouds. This limits particulate absorption as precipitation falls through the air column. Consequently, fresh rain is purified, low in salts, and close to a neutral pH between 5.5 and 6.5. The lack of dissolved minerals makes rain ideal for watering plants and gardens, animal consumption, and numerous light domestic uses.

Potential Contaminants

While falling through the atmosphere, precipitation incorporates various particulate matter present aloft. Sources of such particulates include automobile and industrial exhaust emissions, ash from forest fires and volcanoes, oceanic sea salt sprays near coasts, and microbes from bird and rodent droppings. Waste incinerators, manufacturing facilities, and urban congestion increase airborne pollutants. Pollen and dust also get picked up.

Careful collection and storage prevent most contaminants. Initial dirty runoff from the roof and ground catchment surfaces should be diverted using first flush diverters. Screens filter debris entering storage tanks. Routine maintenance checks and cleaning preserve quality. Proper precautions allow harvesting water to be superior to utility supplies.

Chlorine-Free

Municipal treated tap water contains residual chlorine used to disinfect the supply distribution system. While assuring microbiological purity, chlorination byproducts in drinking water have raised health concerns. The lack of such additives makes rainwater advantageous for those seeking natural, chemical-free water. However, filtration and supplementary disinfection methods may be warranted if testing indicates risks.

Moderate Acidity

The pH value measuring relative acidity or alkalinity falls between 5 and 7 for rainwater, averaging 5.7 worldwide. This mildly acidic level correlates to atmospheric carbon dioxide reacting with water to form weak carbonic acid. The moderate pH suits most purposes without adjustment, unlike some groundwater sources that require conditioning to lower pH levels.

Testing Provides Insight

While fresh precipitation offers very pure water, specific local conditions impact quality. Testing harvested rainwater identifies particulates, organic compounds, pH, bacteria levels, and other parameters. This establishes a benchmark to guide treatment methods and determine suitable uses, whether for drinking, laundry, or garden irrigation. Periodic tests monitor changes and help troubleshoot issues.

Low Dissolved Solids

The low concentration of dissolved solids in rainwater means pure H20 with minimal salts or other inorganic compounds. By comparison, groundwater and surface water pick up significantly more dissolved solids filtering through soil and rock layers. Tap water also gains impurities from municipal treatment and piped distribution systems. For those seeking water with low total dissolved solids, rainwater fulfills this need.

Lack of Hardness Minerals

Hardness in water results from dissolved calcium and magnesium, primarily in the forms of bicarbonate and sulfate ions. As rain falls through the atmosphere, it does not dissolve minerals. This lack of calcium and magnesium ions makes harvested rainwater an exceptionally soft water source. Using rainwater eliminates stiff laundry, foggy shower doors, and scale deposits on fixtures and appliances.

Neutral pH

While carbon dioxide in the atmosphere dissolves slightly in precipitation to form carbonic acid, rainwater remains close to pH neutrality, averaging between 5.5 and 6.5. This moderate acidity-alkalinity balance suits most purposes ranging from drinking and cooking to cleaning and gardening. It falls in the neutral 6-8 pH range appropriate for human uses and ecological purposes without chemical adjustment.

Low Sodium Content

Sodium in drinking water raises health concerns, especially for those predisposed to hypertension or kidney ailments. Sodium salts also react adversely with soaps and detergents. The low sodium concentration in rainwater makes it ideal for those requiring a low-sodium diet or seeking to extend the lifespans of soaps. Unless collected near the ocean, rainwater contains minimal sodium chloride.

No Added Fluoride

While public health officials support municipal water fluoridation to reduce cavities, some oppose this mass medication of supplies. Rainwater provides a fluoride-free alternative. However, families must balance the risks of tooth decay and the benefits of fluoride, considering other sources like toothpaste. Lack of fluoride remains a factor to discuss with dentists.

Softens Laundry

Rainwater's soft quality prevents mineral scale buildup inside washing machines. The low hardness also allows the use of the minimum detergent needed for effective cleaning. Clothes rinsed in soft rainwater come out residue-free, smell fresh, and feel smooth against the skin. Hard municipal water leaves stiff, scratchy fabrics needing extra rinses.

Few Microbes

When rain falls through clean air, it contains little microbial contamination. While possible pathogen risks exist from bird droppings or insects in catchment areas, these get minimized through the diversion of initial roof runoff and proper storage hygiene. Municipal supplies see more contamination from biofilm buildup during distribution. With care, rainwater maintains a high standard of microbial purity.

Nutrient-Rich for Soil

Nitrogen and other nutrients in the atmosphere get incorporated into rain, making it beneficial for soils and plants. Phosphorous and potassium levels are low, but minerals leach from catchment roofs. The light nitrogen fertilizes without risk of groundwater contamination from nitrates. Rainwater's purity prevents saline buildup in the soil.

Factors Influencing Rainwater Harvesting

Successful rainwater collection relies on careful evaluation of key influencing factors. Local precipitation patterns, catchment areas, storage capacity, and household usage rates interrelate to determine feasible harvesting potential. Thorough planning considers climate conditions, infrastructure scale, quality concerns, and usage purpose when engineering a well-designed system. Identifying optimal siting and sizing provides a realistic perspective on how much water can be reasonably harvested to meet needs.

Rainfall Amount and Distribution

The volume and distribution of rainfall impacts collection potential. Arid regions average 10-15 inches annually, but sporadic intense storms provide episodic abundance. Humid areas see frequent light precipitation totaling 30-60 inches yearly. Average rainfall spans 20-45 inches in temperate zones. Study historic monthly and annual totals to understand local norms. Also, track cycles, seasons, and extreme variability. This data informs storage capacity and supply expectations.

Catchment Size

Rainwater harvested depends directly on the catchment size. Larger collection rooftops or ground surfaces generate more supply volume. A 2000-square-foot roof potentially yields over 20,000 gallons annually in most regions. Optimization tools help right-size catchment and storage capacity based on weather data and demand. Careful material selection prevents chemical leaching. Position gutters and pipes to channel flow effectively.

Site Topography and Drainage

Evaluate site contours and drainage patterns when planning tanks, ponds, and water flow. Gravity provides a passive transfer to storage reservoirs. Land sloping toward structures eases rainwater flow into containment. However, permeable sandy soils may require liners to prevent infiltration losses. Divert stormwater runoff around storage to avoid contamination. Consider site low points, swales, and soil permeability.

Storage Capacity

Sufficient storage volume is key to meeting household water needs, especially during dry seasons. Large above-ground tanks or underground cisterns provide capacity but can be expensive. Optimal size depends on catchment area, climate, and usage rate calculations accounting for dry periods. Oversizing risks water stagnation, while undersizing leaves demand unmet. Expand capacity as usage grows.

First Flush Diverters

First flush diverters route initial roof runoff containing accumulated debris and contaminants away from storage capture. However, this bypasses significant volume. Provisioning tanks for bypass when capacity is critically needed allows flexibility. Divert the first 5-15 gallons from roof surfaces, depending on size. Guided by water quality tests, manage the tradeoffs between purity and supply.

Precipitation Monitoring

Active rain gauges measure local precipitation volume and intensity. Data validates historical averages and informs real-time capacity planning and usage. Passive rain barrels with calibration lines offer basic readings but consider dedicated monitoring. Weather stations provide regional trends. Measure seasonal and annual totals. Note intensities for overflow prevention.

Intended Applications

Evaluate primary water applications when planning a system. Gardening and irrigation involve large volumes cyclically. Domestic washing and drinking represent smaller but consistent daily demands. Livestock require reliable watering. Integrating various uses with different cycles ensures adequate supply. Match storage and catchments to total volume requirements.

Usage Monitoring

Actively monitoring household water usage reveals consumption patterns and rates. Use meters to quantify gallons used by appliances, fixtures, irrigation, livestock, and other activities. Conduct periodic audits and inspections to check for leaks. Measuring usage guides adjustments in storage, conservation practices, and expansion priorities to balance supply and actual demand.

Prefiltration and Screens

Leaf screens, roof washers, first flush diverters, and other inlet filtering methods prevent debris and particulates from entering storage. Add sediment filters for ground collection surfaces. Screens keep out insects and animals while allowing good flow rates. Filtering preserves water quality but increases maintenance. Right-size screens for your needs.

Rain Barrels Versus Tanks

Small rain barrels suffice for garden irrigation but limit capacity for household uses. Interconnecting a series expands storage in a modular fashion. Tanks come in a range of vertical and horizontal models from 120 to over 5000 gallons. Weigh costs against durability and volume needs. Tanks may require pump systems, water treatment, and climate control.

Gravity Feed Versus Pumps

Gravity provides a free passive transfer to use points below storage elevation. Overflowing barrel spigots allow gravity hose connections. For pressurized delivery, add pumps with pressure tanks and controllers. Pump systems demand electricity and maintenance but give greater functionality. Analyze tradeoffs between passive gravity feed and pumped systems.

Water Protection and Safety Seal storage against sunlight, pests, algae, and airborne contaminants. Exclude children from access with locked covers. Locate components safely given water weights. Label color-coded pipes. Check for bacterial or chemical contamination. Implement maintenance procedures to keep the system sanitary and safe. Monitor conditions to prevent health hazards.

Rain Catchment Material

Catchment surface materials impact water quality. Collecting from roofing like asphalt, painted metal, and some treated wood introduces dissolved compounds. Better choices include stainless steel, untreated wood shakes, and certain composites or polymers. Evaluate potential leaching from new materials using lab analysis. Consider longevity and sustainability.

CHAPTER 3
LEGAL CONSIDERATIONS

Jenna was excited to set up a rainwater harvesting system on her new homestead. She had researched various designs and was ready to install gutters and a storage tank. However, when she mentioned her plans to a friend, he cautioned her about seeking necessary permits first. Perplexed, Jenna did more research and was surprised to discover the complex web of regulations governing rainwater collection in her state. She realized ignoring laws could land her in trouble. This chapter aims to help homesteaders like Jenna navigate the legal landscape of rainwater harvesting through education and adhering to local statutes.

Overview of Regulations Worldwide

Rainwater harvesting policies vary significantly around the world. Some countries and regions actively promote and subsidize it, while others impose taxes or outright bans. Understanding prevailing laws is crucial before establishing any system.

Climate and Water Availability

Regions with abundant rainfall face fewer restrictions compared to water-scarce areas. Coastal regions usually have more lenient norms given their proximity to oceans as a readily available water source. However, inland and drought-prone locales regulate rainwater usage more stringently to prevent over-extraction from the hydrologic cycle and depletion of limited freshwater supplies. Areas experiencing chronic water shortages have understandably stricter laws around rainwater harvesting aimed at conserving water resources for community needs.

Water-rich coastal cities commonly allow unfettered collection and usage of rainwater. For example, Seattle and Dublin face very minimal legal barriers for homeowners and businesses to harvest, store, and utilize rainwater runoff from their rooftops without requiring permits. This relaxed approach reflects ample local water availability, moderate year-round rainfall, and strategic emphasis on decentralizing stormwater management through low-impact development techniques, including rainwater harvesting wherever feasible.

Land Ownership Types

Public lands fall under jurisdictional oversight, whereas privately-held properties enjoy more freedom depending on state laws. Some communities prohibit non-potable uses like irrigation on communal territories but allow private residential catchments. This distinction stems from the consideration that individuals have higher entitlement and discretion over activities conducted fully within the confines of privately owned property compared to public lands managed for community access and needs.

State and national forests, parks, wildlife refuges, etc., constitute public lands where stricter control exists over any alteration to natural hydrological processes like large-scale rainwater harvesting, which could impact streams, wetlands, or aquifer recharge within their territories. On the other hand, private agricultural or residential plots see more lax rules, assuming systems are moderate in size and follow best practices to avoid disturbing surrounding land and water bodies.

Domestic vs Commercial Usage

Household-scale projects receive relaxed consideration relative to larger installations meant for business or agriculture. Volume thresholds may apply as commercial harvesting impacts local aquifers more substantially due to higher collection and usage rates from vast hardstanding surfaces like multiple building rooftops, parking lots, etc. Some states restrict non-potable business usage outright due to this increased burden and preference given to potable water needs of municipal sources.

Specifically, cities like Tucson and Phoenix prohibit commercial harvesting of more than 10,000 gallons per year due to scarcity issues despite allowing unconditional private harvesting of under 3,000 gallons annually. This balanced approach factors in varied water demand and impact from different sectors. Private homes utilizing rainwater for non-essential uses have a relatively lower impact than agricultural or industrial operations dependent on sizable non-potable supplies for process and irrigation water needs year-round.

Rainwater Rights

A few arid Western states assert public ownership of precipitation, citing historical water laws inherited from the prior appropriation doctrine that prevailed during settlement times. However, most coastal and eastern regions consider rainfall a private property extension like trees or crops that landowners can freely collect or direct from their plots. Rights vary significantly in this regard between states depending on their legal traditions and water circumstances, which molded salient statutes.

In parts of Colorado and Wyoming following the prior allocation system, any diversion or storage of rainwater beyond a small amount requires a water right like stream water, even if occurring entirely on private property. This interpretation of rainwater as a public resource alludes to its criticality in drought-prone areas. On the contrary, states like Florida and Maine treat precipitation falling on rooftops or land as an inherent private right similar to the collection of other landscape products like fruit or timber within property confines. Most homeowners do not need explicit permits here to harvest rainwater.

Drinking Water Standards

Most industrialized nations subject potable rainfed supplies to treatment and testing under public health codes to ensure safety for drinking purposes. Developing countries occasionally waive these for small self-supplied homestead systems supported by cheaper collection occurring at the household scale. However, reliable treatment becomes important for public drinking water systems dependent on rainwater.

Specifically, the US EPA mandates microbial standards through the Surface Water Treatment Rules for any utility utilizing surface water sources, which now, in many states, include cisterns supplied by rainwater due to recognition of its similarities to open water sources. This ensures disinfection and monitoring comparable to standards for well or creek water to eliminate risks of gastroenteritis and other illnesses from untreated supplies. Separately, Australia and many European nations rely on AS/NZS 4020 and DIN EN 1717, respectively, to certify water treatment solutions and plumbing installations involving rainwater usage for potable intake are safeguarded.

Untreated non-potable collection meanwhile faces fewer directives globally for uses like irrigation, car washing, toilet flushing, etc., provided systems don't cross-connect with drinking supplies by following air gap principles. This lowers regulatory burdens without compromising public health, as rainwater is not intended for internal consumption in such cases. The same best practices centralized water treatment upholds for delivering safe water to taps are thereby unnecessary.

Building Codes

International plumbing standards exist, but enforcement quality varies substantially by region. Integrating rainwater catchment components during new construction usually meets less bureaucratic resistance than retrofitting occupied structures if they meet base code requirements. This preference considers retrofits require altering plumbing infrastructure, which may impact housing integrity if shoddily implemented, whereas greenfield projects face no such constraints in optimizing for rainwater reuse through dutiful integration during initial planning, approvals, and erection itself.

Specifically, graywater plumbing retrofits in occupied housing need careful review and permitting to ensure resident health and safety through considerations like preventing crossflows, backpressure effects, and unlawful cross-connections into potable systems. Meanwhile, the International Green Construction Code offers an optional regulatory framework many localities have adopted, which eases new builds to voluntarily incorporate sustainable features like on-site collection, conveyance, and end-uses of precipitation, which impacts approval timelines minimally when designed prudently from the outset. Such foresighted planning deserves relaxed treatment vis-a-vis reactive upgrades to existing structures.

Environmental Impact Regulations

Natural runoff patterns and rates must remain largely unchanged to prevent disruption of aquatic habitats through alterations in flow regimes or water quality, which large-scale diversions could potentially induce. Therefore, most jurisdictions require projects exceeding a minimum threshold, like 5000 gallons of storage or 2500 square feet catchment area, to undergo environmental impact assessments, ensuring regional hydrology and ecology remain safeguarded per environmental laws and best practices. Assessment criteria typically consider catchment area sizing relative to parcel/watershed size, runoff detention integration, and maintenance of environmental flows in adjacent watercourses and their floodplains to predevelopment conditions through controlled release provisions.

For example, Canada's Fisheries Act stipulates rainwater harvesting projects should not impede fish passage or spawning activities in any way, such as through improper storage releases affecting hydroperiod fluctuations tolerated by indigenous species. This necessitates ecosystems receive an equivalent level of protection from hydrological modifications, whether through rainwater harvesting, irrigation withdrawal, or other diversions. Environmental regulations thereby strive for water balance through a combination of low-impact infrastructure with sensitive release mechanisms as a condition for project approvals larger than household scale.

Downspout Disconnection Laws

Some municipalities and water agencies offer incentives for delinking roof drainage pipes from storm sewers via passively directing runoff onto permeable areas instead of combined infrastructure to benefit both water quality and supply augmentation objectives. Downspout re-routing effectively contributes to reduced pollution and overflow risks to local water bodies from sewer lines overloaded during intense storms, transporting contaminants directly from streets and gutters into receiving streams untreated. Simultaneously, infiltrating roof runoff replenishes subsurface stores and buffers peak flows - outcomes rainwater harvesting exponentially builds upon by containing redirected flows for multiple indoor and outdoor non-potable applications with the convenience of an alternate local source independent of potable treatment and conveyance infrastructure.

Several regional EPA programs recognize downspout disconnection and routing to private landscapes as low-effort household-scale projects perfectly aligned with sustainable drainage objectives laid out in their municipal separate storm sewer permits issued to cities and counties. Participating properties can even become eligible for cost-share grants offsetting modest retrofit costs to voluntarily help municipal separate storm sewer system (MS4) communities progressively achieve pollutant load reductions mandated in their stormwater management programs through collective decentralized, distributed actions.

Data Reporting Requirements

A few states inventory all surface and subsurface water usage to monitor supply and demand trends for policymaking purposes such as Colorado, Texas, and Oregon in the western U.S. Self-supplied homeowners may need to declare catchment system details, including collection area and storage capacities during permitting processes, though ongoing reporting burdens often exclude small residential rainwater harvesting which comprises an imperceptible portion of total usage.

Larger or commercial-scale projects may face additional annual compliance requirements involving the submission of meter readings, as-built drawings, and occasional inspections, depending on state mandates. For example, Texas law designates all rainwater harvesting as groundwater usage, necessitating reports to the groundwater conservation district of the county. However, catchments under 3 acre-feet (972,000 gallons) receive streamlined treatment with one-time registration sufficient.

Certain drought-prone Australian states like Queensland have also instituted registration of all domestic and commercial rainwater tanks above a 5000-liter threshold to inventory diffuse non-potable demand subtracted from treatment plant load. While imposing paperwork, such measures furnish a broader understanding of water flows to guide subsequent policy and help prioritize augmentation initiatives. Maintaining current stock inventories supports equitable management, especially during shortage periods necessitating demand curbs.

Monitoring Water Quality

As natural runoff can accumulate contaminants from roofing materials, surrounding soils, and atmospheric deposition, most regulatory agencies recommend basic microbial and chemical testing of stored rainwater periodically or before potable consumption from household cisterns. Routine screening helps ensure that harvested water continues to meet intended usage guidelines.

For example, Australian guidelines stipulate testing rainwater tanks supplying drinking water applications annually at minimum for E. coli and turbidity. Additionally, larger commercial catchments intended to partly supply non-potable demands of facilities like greenhouses or sports fields undergo quarterly physicochemical analysis of pH, iron, manganese, and hardness beyond microbiological parameters to characterize suitability for various end uses. Proper maintenance practices help preserve captured water quality between sampling intervals by minimizing the ingress of debris and undesirable constituents over time.

Equity and Affordability Considerations

For developing regions or impoverished rural communities dependent on groundwater wells or centralized supplies, rainwater harvesting holds untapped potential for equitably meeting needs through distributed decentralized systems installed at household or community scale at low per capita costs relative to treatment plants or pipelines. However, onerous regulatory restrictions and compliance costs could preclude its uptake where most needed.

While rules aim to uphold water management objectives, policymakers must weigh the impacts on vulnerable populations seeking self-supply alternatives. Certain exemptions or incentives could motivate harvesting, partly addressing equity by boosting affordable access for those currently underserved or spending excessive shares of income procuring basic needs. Achieving equitable supply security represents the overarching sustainability goal, which may occasionally necessitate relaxed enforcement specially catered to poverty circumstances unique to developing nation contexts.

Integrating Future Policy Responses

As climate change projections forecast worsening water insecurity globally, forward-thinking policy evolves, factoring in unprecedented challenges. Australia's Capital Region is currently debating legislating mandatory residential rainwater tanks, recognizing their ancillary community flood protection benefits besides augmenting supplies. Long-term predictability helps stakeholders gradually adjust to emerging problems proactively instead of reactively struggling with unforeseen consequences of deteriorating hydrometeorological patterns.

Rainwater laws worldwide must thoughtfully incorporate climate adaptation responses sustainably, securing water resources' status quo as precipitation regimes transition. Achieving such a vision demands continuous policy evaluation and reformulation incorporating the latest science while considering on-ground impacts and feedback. Regulation impact assessments coupled with open stakeholder consultations prove indispensable in making rainwater harvesting frameworks more resilient and future-proof over the long run.

Seeking Permissions and Adhering to Local Laws

"I was surprised to discover my state requires rainwater harvesters to obtain annual operating permits," shared Jenna. "Luckily, the application process was simple." Homesteaders must research prevailing regulations for their property location, zone, and intended usage. Contacting county officials proves invaluable. Rules often aim to balance rights and responsibilities.

Check with Local Regulatory Agencies

Jenna consulted county water managers, who informed current laws and the permitting pathway. Similar first stops include health or public works departments based on project scope. Officials clarify mandate applicability case-by-case. Consulting the administrators enacting statutes represents the surest route to compliance.

Most regulatory bodies maintain websites compiling jurisdiction-wide ordinances searchable by topic. Jenna benefited from easily accessing rainwater harvesting code chapters applicable to her county. Staff also fielded questions resolving ambiguities. Such upfront due diligence saves future headaches and ensures that proposed systems pose no unintended issues.

Determine if an Operating Permit Is Needed

Permits exempt small household systems below predefined size thresholds. Jenna's single 5000-gallon tank fell under limits. Larger commercial projects typically require permits along with engineered designs. Careful measurement against quantitative criteria streamlines the approval process.

For instance, Mississippi enables unfettered rainwater harvesting for non-potable uses from catchment surfaces below 500 square feet and storage below 5000 gallons without oversight. Larger collection necessitates permits, considering increased demand and impact. Proper classification according to established numeric standards avoids superfluous paperwork.

Adhere to Setback, Parcel Size, and Other Stipulations

Jenna confirmed her half-acre lot met land area standards. Setbacks from property lines ensured the preventing of runoff issues on neighbors. Special care upheld rules in high water tables or flood-prone locations to avoid contamination. Careful document review prevents costly post-installation adjustments.

For example, pipes releasing stored rainwater to the landscape must discharge at least 10 feet from any septic system components in many counties to preclude hydraulic interference or crossflows per health codes. Conforming to mapping verifies the intended layout and poses no conflict.

Implement Best Practices for Health and Safety

Jenna opted for retrofitting with approved, non-toxic materials. She included easy access ports and clean-outs and installed fine screens covering gutters and inlet pipes. Dual plumbing segregated potable supply. Clear signage labeled non-potable rainwater and its end use, like irrigation or appliance feeds. Adhering to technical standards protects users and property.

Building codes typically necessitate screening out debris at gutters and pipes to maintain captured water quality. Safety ports every so many linear feet facilitate flushing sediments periodically without draining entire tanks. Isolating drinking water ingress through air gaps prevents accidental cross-connections.

Check for Inspection Requirements and Timelines

Jenna's collection system underwent a final inspection before operation. Annual compliance checks upheld quality. Some states require licensed contractors for larger commercial installations with stricter oversight conditions. Timely permit renewals prevent penalties. Compliance protects future usage rights and community health.

Certain municipalities demand contractors certified in rainwater catchment system installation. Inspections particularly scrutinize materials, treatment, and proper separation from potable lines. Regular maintenance inspections then uphold initial standards yearly.

Community Rainwater Harvesting Programs

"The county offers rebates and design guidance for cooperative projects benefitting neighborhoods or farms," mentioned Jenna. Larger scale efforts complement community water management plans through conjunctive use of local supplies. Government incentives increase adoption rates.

Some local governments subsidize the bulk purchases of components for neighborhoods to collaboratively install shared tanks. Partnerships between agencies and community groups also produce educational videos enabling do-it-yourself harvesting within statutes.

Continuous Education and Responsible Usage

Jenna remains proactive by attending public awareness sessions. She shares expertise in helping others harvest ethically within prevailing statutes. Responsibility and conservation-focused usage ensure regulatory compliance and community acceptance of the practice long-term. Education builds durable stakeholder understanding.

Most jurisdictions provide annual training covering installation tips, maintenance best practices, and regulation updates. Workshops help harvesters network while strengthening personal understanding, reinforcing statutory spirit and intent behind various measures considering all perspectives.

Operating Permit Application Process

Most permitting departments publish simple one- or two-page forms listing basic system and applicant information. Jenna faced minimal hassle providing details like storage volume, catchment area, and intended end uses. Streamlined paperwork encourages participation.

Some applications require more specifics, such as site maps denoting setbacks, plumbing schematics, and a basic operation and maintenance plan. Sketches clarify layouts, while maintenance plans reassure regulators systems receive care upholding functions and water quality over the lifetime.

Permit Review Timeline

Regulators commit to reviewing rainwater harvesting permits within a fixed number of days, usually 30, based on completeness. Incomplete applications result in requests for missing elements, tacking on review periods, which frustrates harvesters. Frontloading submittals avoid drawn-out wait times.

Jenna's fully documented form underwent prompt approval within two weeks as township staff faced no backlog. Expedited treatment incentivizes voluntarily working within statutory frameworks, sustaining productive stakeholder relationships.

Recertification Requirements

Most permits require periodic renewal, commonly every five years, accompanied by an inspection verifying adherence to original conditions and codes, ensuring quality stewardship over the long term. Periodic oversight maintains mutual accountability.

Renewal checklists guide harvesters, refreshingly updating documentation to streamline compliance. Jenna faces recertification inspection involving minor maintenance like gutter cleaning every half-decade upholding her operating permit.

Enforcement Protocols

While authorities rely primarily on education addressing non-permitted harvesting, reckless systems are threatening health warrant penalties. Proactive harvesters face leniency, while reckless actors face fines. Mitigating hazards necessitates equitable yet judicious administration.

Officials issue notices to unauthorized collectors demanding applications or upgrades within 30 days prior to prosecution, preserving harvesters' rights to remedy issues cooperatively. Willful violations, e.g., endangering neighbors, warrant court-mandated remediation.

Appeals Process

Rejected or penalized parties maintain a grace period for voicing reconsideration requests to independent review boards. Natural justice outweighs rigid bureaucracy, fostering consensus. Win-win solutions balance flexibility with accountability.

Jenna surfaces, hoping to quadruple her tank size. Board members visit endorsing expansion with buffer zone plantings addressing neighbors' oversaturation concerns, amicably resolving the dispute.

CHAPTER 4
ASSESSING YOUR RAINWATER POTENTIAL

A homesteader stood on the edge of their property, admiring the rolling hills and valleys that surrounded their home. As they gazed at the landscape, their mind began to turn towards sustainability and self-sufficiency. They had been considering installing a rainwater collection system to help meet their household and agricultural water needs, but first, they knew they needed to properly assess the potential of their land. This chapter will walk through the key steps and factors to evaluate when gauging a property's rainwater harvesting potential.

Calculating Roof Surface Area

The roof is essentially the "collection area" for any rainwater harvesting system. A larger roof translates to more rainwater capture potential. Therefore, the first step is to accurately determine the total roof surface area that can be utilized for rainwater collection. This assessment forms the foundation for appropriately sizing all other system components.

Measuring the Roof Dimensions

The homesteader began by pulling out a 25-foot measuring tape and notes. Starting at one corner of the house, they extended the tape along the eave to where it met the next wall or angle. This length was recorded. They then paced off and measured the width of that section at several points to account for any variations, noting the minimum and maximum figures.

For the main gabled roof section that covered most of the house, a length of 35 feet along the eave and an average width of 22 feet was observed. Moving towards the back of the house, a small addition was evident by its change in materials. This rectangular area measured 12 feet by 10 feet and had its own independent roof.

Converting to Square Feet

With all the linear measurements taken, it was time to convert to square feet by multiplying length by width. For the primary gabled section, the calculation was 35 feet x 22 feet = 770 square feet. The small rear addition worked out to 12 feet x 10 feet = 120 square feet.

Rather than doing math in their head, the homesteader opened their notebook and jotted down each shape measured along with the dimensions. In separate columns, they wrote the multiplication and the resulting square footage. This provided an initial subtotal and allowed methodically working through each section without losing track of numbers as more complex roof geometries were encountered.

Accounting for Obstructions

Examining the roof, two small vent pipes poking through were observed. Using a tape measure, their circumference was found to be roughly 1 foot each. Approximating them as circles, the area of each was calculated as πr^2. For a 1-foot radius, this equaled around 3 square feet per vent pipe.

In addition to the vents, the homesteader spotted a newer solar panel array mounted towards the rear. They paced off its length and width, finding it encompassed a 4-foot by 8-foot space. Double checking with a tape measure confirmed the 32-square-foot area is no longer suitable for rain collection.

Including Outbuildings

On the property was an adjoining woodshed and tool shop. Both had simple gabled roofs but would contribute extra potential. Measuring tape in hand, the homesteader went to inspect them.

The small tool shop proved rectangular, coming in at 10 feet by 15 feet. Simple multiplication provided the 150 square feet. Nearby, the woodshed occupied a 12-foot by 8-foot footprint for yet another 96 square feet. Careful recording in the notebook preserved all figures.

Factoring Sloped Roof Efficiency

One consideration with sloped roofs is that not the entire surface area acts as an effective collection region. Only the projection of the sloped area onto a flat plane can realistically harvest rainfall.

To account for this, the homesteader referred to roof pitch calculation guides. The main house had standard 3:12 slope shingles. Looking up this ratio, it indicated only 45% of the total sloped surface collects water like a flat plane would. Therefore, only 45% of the 770 square foot main roof, or 346.5 square feet, should be counted as an effective collection area.

Documenting the Calculations

As areas and obstructions were tallied through systematic surveying, the homesteader took care to neatly log everything on graph paper for clarity. Each roof component received a numbered section on the paper. Corresponding notes provided the linear measurements, calculations, partial area adjustments, and final square footage figures.

Color coding was also added to distinguish sections like the main house in blue and the outbuildings in green. The key was maintaining an organized workbook that could serve as a planning guide or reference during future inspections and improvements.

Re-evaluation Over Time

The homesteader was aware homeowners commonly upgraded or expanded parts of their properties over the decades. To keep their roof area data accurate, periodic rechecks were deemed important.

A note was made on the graph paper to schedule the first re-evaluation in 5 years, assuming no major changes occurred sooner. This would coincide with evaluating the whole rainwater system and making revisions as needed. Additional verifications could then follow every subsequent half decade to capture any roof alterations.

Secondary Collection Areas

Beyond only the roofing surfaces, other flat, industrial areas held potential if retrofitted properly. Scanning the landscape, one location stood out.

A large gravel parking pad measured approximately 20 feet by 30 feet. Though not currently set up to funnel water anywhere, its solid, impermeable nature lent itself well to the collection. Some thought was given to how piping and a simple berm could be incorporated to tap into this extra 600 square feet.

Translating to Expected Yields

Now that collection dimensions were quantified, putting the numbers in the context of typical yields helped set realistic production goals. Regional rainfall records showed average annual amounts ranged from 35-45 inches. The research found that 1 square foot of roof produced around 30 gallons annually in these conditions. Therefore, the homesteader estimated their 1,120 square foot system equated to about 33,600 gallons per year before accounting for losses.

Understanding Rainfall Patterns and Calculations

While the roof measurements revealed the homestead's collection capabilities, the homesteader knew rainfall amounts and their distribution through the year highly impacted yields. A close investigation of historical records provided deeper insights to appropriately size the system.

Consulting Historical Rainfall Records

The county extension office housed a library of climatological data. In the weather section, yearly rainfall totals dating back decades caught the homesteader's eye. Pulling valuable books from the shelves, they began annotating tables chronicling monthly and annual precipitation since the 1900s recorded at the nearest weather station. Averages were broken out by decade to observe fluctuations over generations. The comprehensive archives proved a treasure trove of context for current conditions.

Identifying Nearby Weather Stations

Back at home, topographic maps revealed three stations within 20 miles. To determine the most fitting, each location was analyzed. The nearest site sat in a low-lying valley, potentially yielding lower totals due to rain shadowing. Another fell along a major lake, risking artificially high readings. The third laid at a similar elevation and several miles inland, seeming most analogous. Cross-referencing with rainfall contours supported selecting this representative station.

Analyzing Monthly and Seasonal Trends

Examination of the records from the chosen station uncovered cycles. Winters featured the most rain and snow. Late spring brought occasional powerful storms. Summers generally saw the least moisture with periodic drought risks. A line graph was sketched plotting average precipitation by month, visually depicting this ebb and flow. Dry periods could now be anticipated and prepared for through intelligent storage capacity.

Calculating Average Rainfall

Averaging rainfall data tabulated in the station publications provided perspective. Yearly values ranged from 29 to 52 inches, with an arithmetic mean of 38 inches over 30 years.

To determine the normal precipitation quantity by month, the total inches each received were added up and divided by 12. This revealed a general distribution of around 3 inches from November to February and 2-3 inches during other seasons.

Factoring in Climate Change Impacts

Comparing graphs of local rainfall trends against global temperature curves indicated conditions had become wetter overall by roughly 5% in the last few decades. More extreme events also seemed common. Contact with a climatologist ensured these expected changes were incorporated into storage sizing rather than relying on potentially outdated historical averages alone. Forward-looking planning prepared for climate unpredictability.

Understanding Rainfall Depth and Intensity

Digging deeper into the daily station logs, most storms dropped 0.1-1 inch. A few powerful systems unmatched the area's annual average in only an hour. This helped contextualize intense short bursts filling tanks rapidly versus drawn-out gentle rains. Impacts on retention times, runoff potentials, and first-flush diverters became clearer with these intensity insights factored in.

Differentiating between Rain and Snow

Given winter snowfall averages also shaped hydrology patterns, freezing precipitation needed to separate out from rainfall stats. Through cross-referencing regional isohyet AI maps with winter temperature bands, estimations of 15-25 inches of typical annual snow water equivalent emerged. This liquid quantity was subtracted from total precipitation figures for an accurate picture of harvestable liquid volumes.

Identifying Errors and Anomalies

Outliers were examined for potential recording mistakes, like a reported "2-inch rainstorm" alongside snow-denoted days. Flagging these one-offs avoided skewing expectations.

Periodic climate station maintenance checks uncovered prior sensor issues potentially impacting data quality in isolated years as well. Only verified measurements formed the basis of long-term norms.

Including Location Notes

To preserve crucial geographical context, notes accompanied rainfall data mentions. Details like the weather station residing in a floodplain provided insight into how location impacted its hydrology when referencing figures years later. Similarly, naming the specific county and nearest cities documented place-based influences beyond raw numbers alone.

Potential Storage Needs

Proper sizing of rainwater storage components was paramount for the homesteader to fully benefit from their collection efforts year-round. A meticulous study of property needs and rainfall patterns has now laid the foundation for intelligent capacity calculations.

Listing Current Water Uses

The homesteader gathered all household members to brainstorm their daily water activities. A whiteboard soon filled with uses like cooking, drinking, bathing, laundry, and cleaning listed from most to least essential. Outside, gardening and crops topped the chart. Through discussions, tasks involving water, such as vegetable plots, berry patches, orchard trees, and ornamental plantings, were pinpointed. More frivolous applications involving recreation ranked at the bottom.

Identifying Peak Demand Periods

Cross-referencing water needs with local weather patterns showed summer months required the greatest storage volumes. Without rain for over two months, supplying irrigation drainage became pivotal. Extensive vegetable harvesting from June through August demanded large volumes. Simultaneously, ornamental garden upkeep also peaked. If rains failed, stored water ensured continued productivity and household welfare.

Calculating Daily Water Usage

To gauge day-to-day requirements, the homesteader logged flows from indoor taps and spigots for a week. Every toilet flush, shower, load of dishes, and gardening session received a tally. Indoor usage averaged around 40 gallons per person daily. Outdoors and peak watering nights exceeded 150 gallons as drip lines, and soakers ran through the evening. Combining all data revealed current consumption reached up to 1,000 gallons on the most intensive landscaping days.

Factoring in Planned Expansions

In future years, the homesteader hoped to expand poultry and egg operations. Accounting for these aims showed each chicken required 1 gallon of water per day. Preliminary plans included constructing a small barn to house 50 birds. At 50 gallons daily just for poultry, incorporation into the storage sizing projections became prudent rather than experiencing shortages down the line.

Coordinating with Rainfall Patterns

Monthly rainfall averages provided context for sizeable storage. From June to September, as little as 1-3 inches typically fell each month, per historical data. During these rain-scarce summer periods, supplying non-potable needs solely from rainwater collection dictated minimum sizing. Ensuring adequate reserves prevented the demise of crops or purchasing supplementary water.

Calculating Required Gallons

Tallying typical summer demands of 1,000 gallons per day for two months equaled 60,000 gallons required if relying solely on stored water. Adding a 5,000-gallon buffer and the 50 chickens equating to 25,000 more gallons, preliminary minimum storage needs were estimated at 90,000 gallons. Sizes available were cross-checked against this baseline. Storage reservoirs held thousands more than projected working needs, allowing for hazier periods beyond average weather patterns as an extra safeguard for self-sufficiency.

Allowing for Unpredictability

While historical patterns provided guidelines, climate fluctuations meant less reliable predictions further into the future. Drought risk involves deeper uncertainties. Therefore, to play it safe, another 10,000-gallon buffer was added as a conservative measure. This brought the total recommended storage up to the recommended maximum of 100,000 gallons before selecting a tank.

Factoring in Input Rates

Given the 1,100 square foot roof, rainfall intensity logs showed a typical 1-inch rainstorm delivered over 800 gallons within an hour. Extrapolating recharge rates out, 100,000-gallon storage replenishing 10-20% after major storms suggested maintaining healthy reserves through all but extraordinarily dry seasons, barring equipment faults.

Selecting Tank Sizing

Three galvanized steel tanks between 85,000-105,000 gallons emerged as suitable options. The 105,000-gallon reservoir would provide ample redundancy, allowing for supply in low rainfall decades. Its $4,000 cost fitted comfortably in the homesteader's budget when factoring in the long service life of properly sited and maintained tanks.

Prioritizing Potable Supply

While crops took priority for irrigation, household water needs topped the list. By front-loading potable storage capacity, a reliable 8,000-gallon buffer ensured domestic health even during drought. This thorough storage sizing methodology established a framework granting the homesteader complete confidence in their rainwater independence for both current and future water security. The system design phase could now commence in full.

CHAPTER 5
COMPONENTS OF A RAINWATER HARVESTING SYSTEM

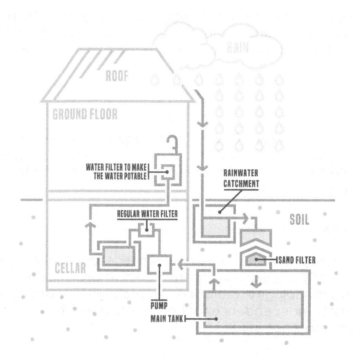

Just as the various organs in the human body work together in harmony to sustain life, each component of a rainwater harvesting system plays an integral role in efficiently collecting and storing nature's free gift of fresh water. When designed and installed correctly based on the principles of watershed hydrology, a rainwater harvesting system can keep supplying non-potable water needs for homes, gardens, farms, and more; this chapter explores each key part that forms the whole, from capture to storage to distribution and beyond.

Catchment Surface: Roofs and Beyond

A suitable catchment surface is the starting point of any rainwater harvesting system.

Catchment Material - Roofs fabricated from corrosion-resistant materials like galvanized iron, copper, or concrete are most suitable for rainwater catchment. Properly installed and regularly maintained roof catchments can last for decades with minimal deterioration of water quality.

Roof Shape - Gabled or hipped roofs are preferable to flat or almost flat roofs as they allow water to flow easily into gutters. The steeper the pitch, the better the water flow. Standing seams on metal roofs can act as mini-gutters to channel water flow.

Roof Size - Larger roof catchment areas translate to more rainwater harvest potential. Each square foot of roof can yield 15-30 gallons of water annually, **depending** on rainfall patterns. Sizing a storage tank accordingly is essential.

Beyond Roofs - Concrete driveways, sidewalks, open hardstand areas, and specially constructed catchment surfaces can be utilized where appropriate. Pervious pavers allow rainfall infiltration while still supporting overflow to the harvesting system. Grass and soil-filtered rainwater from open areas require more treatment before storage. Protecting catchment surfaces from debris, bird droppings, and vegetation is important to water quality.

Roof Cleaning and Maintenance

Keeping roof catchment areas clean ensures optimal water yield and quality. Fall leaf removal prevents clogging of gutters and downspouts. Pressure washing roof surfaces 2-4 times a year removes dust, algae, and bird/animal droppings. Corrosion-resistant coatings on metal roofs prolong service life. Regular inspections identify repairs quickly, like fixing leaks, resealing joints, or replacing damaged sections. Proper drainage away from the house foundation prevents moisture intrusion risks.

Site Design Factors

Planning catchment placement factors in local slope, wind patterns, and vegetation. Directing downspouts away from building foundations using underground pipes avoids structural issues. Position them to discharge above-ground cisterns for gravity-fed storage. Clearing trees and pruning branches nearby up available roof space while reducing debris. Catchment design aligns with lot layout, utility access, landscaping, and aesthetic preferences.

Gutters and Downspouts

The conveyance system delivering runoff from catchment areas into storage utilizes gutters and downspouts.

Gutter Material - Aluminum and galvanized steel models withstand weathering best. Dark-colored vinyl or resin gutters absorb heat, risking leaks.

Installation - Splash blocks below downspouts divert flows away from foundations. Proper pitch and leak-proof joints channel water smoothly. Downspout extensions empty into pipes or channels. Clear gutter surfaces aid drainage.

Gutter Sizing

Width, depth, and placement relate to roof size and rainfall intensity. Undersized gutters overflow with each storm, while oversized ones cost more. Guide charts assist in sizing for various roof areas, pitches, and climates. Continuous gutters work best, limiting leakage-prone seams. Adding gutter guards staves off debris build-up that causes clogs and backups. Cleanouts ensure maintenance accessibility.

Gutter Maintenance

Seasonal cleaning keeps internal lines clear, whether done with poles and scoops or pressure washing. Checking for damage and repair needs occurs regularly to fend off leaks that steal potential harvest. Painting or treating bare galvanized surfaces preempts corrosion. Adjusting sagging sections or re-attaching loose ones remedies drainage issues. Well-maintained gutters and downspouts serve the system's hydraulics dependably for many rainy seasons.

Alternative Capture Methods

When rooftop systems aren't plausible, curb cuts in drip lines transport flows or French drains underneath porous pavements intercept subsurface runoff. Naturalized swales trap and distribute rainfed into storage via pipes or permeable soils. Crafted catchment areas can comprise any surface permitting gravity-fed transport to cisterns and tanks. Imagination and site conditions together find answers.

First Flush Systems

The initial runoff from catchment areas contains higher sediment loads requiring diversion away from storage.

- **Basic Deflectors** direct first flushes off rooftops using gutters filled with rocks.
- **Downspout Diverters** contain a set volume for rinsing before switching flows into cisterns.
- **Internal Tank Diversions** utilize overflows above intake lines.
- **Mechanized Valves** open only after storms.
- **Sized for Design Storms**, they balance flushing needs with capturing the next rains efficiently for tanks and plants.

Deflector Options

Rooftop rocks, boards, or commercial products fitting inside gutters pond rainfall allowing sediments to wash through before capturing the cleaner remainder. Bucket-like diverters under downspouts contain first rinses that drain elsewhere. Internal tank baffles release an initial overflow. Adjusting size factors in local precipitation patterns and catchment surfaces. Less diversion allows more harvesting but risks dirtier water.

Proper Sizing

Deflector capacities are subject to sizing formulas tied to the roof area, runoff coefficient, rainfall intensity, and required detention duration. Undersized units inadequately flush, while oversized ones squander yield. Peak flow charts assist in right-sizing. Flush volumes between 15-30 gallons per 1,000 square feet applied in under 6 hours remove most floatable solids and organic matter.

Routine Maintenance

Regularly emptied deflectors and diverters preclude clogging and spillover malfunctions. Internal overflows require infrequent dredging of settled debris. Monitoring first flush mechanisms during storms ensures proper functioning that safeguards water quality entering storage. Minimal sediment load remains beneficial for garden soils.

Storage Tanks and Barrels

Rain harvest storage vessels conserve the commodity until households, businesses, or landscapes put it to use. **Tank Types** vary from buried concrete cisterns to plastic, galvanized steel, or fiberglass holding large volumes to smaller, inexpensive above-ground drums. Underground drainage perforated tanks also qualify.

Sizing Storage Capacity

Factors involve catchment area, annual rainfall patterns, intended end uses and any backup drinking water needs. Guidelines tabulate minimum recommended gallons based on roof square footage, indicating adequate self-sufficiency during dry spells. Example: A 1,500 square foot roof in 30 inches/year rainfall locale merits a 1,000-2,000-gallon tank. Oversizing expands future uses.

Tank Location Guidelines

Above-ground versions, by definition, lie openly accessible, while buried tanks require factory-sealed risers or manholes for inspection and cleaning. Foundations support heavy units, with underground contraptions necessitating structural engineering oversight. Placement observes setbacks to minimize tampering risks and avoid seasonal floodwaters.

Materials of Construction

Structural strength, durability against typical elements, contaminant resistance, toxicity concerns, and, of course, cost factors in selecting material types of manufacture. Examples are concrete (permanent structures requiring formwork), galvanized steel or modern rotomolded cross-linked polyethylene, popular for portable containers. Wood rapidly deteriorates from prolonged water saturation and isn't recommended.

Installation and Anchoring

Whether dug or delivered, proper installation methods ensure design lives from storage vessels. Underground structures receive bedding, backfill, and waterproofing. Lifting eyes, straps and reinforced walls aid heavy concrete tank placement into excavations. Setting all tanks on solid, level foundations prevents settling issues. Anchoring above-ground barrels secures them against uplift and movement from winds or ground sloshing. Underground systems couple intake/outlet plumbing for functionality.

Concrete Tank Construction

Reinforced pre-cast units arrive ready for burial, but contractors also pour tanks in trenches using formed panels. Integrated first-flush overflows simplify designs. Vents, access risers, and screened lids allow inspection/repairs without draining. Water-blocking membranes are laid between the foundation and walls/base, plus on the exterior, seal the reservoir.

Steel Tank Installation

Upright models align vertical seams for strength. Per plans, plumbing fixtures get installed through access points sealed with gaskets. Backfill compacts firmly around while leaving vent/inlet pipes accessible above grade. Painted finishes on bare steel with slow outer corrosion with occasional touch-ups.

Maintenance

Routine checks for leaks, drainage issues, or structural settling ensure long lifespans. Cleaning out sediment accumulations from first flush diversions or tank bottoms occurs every 5-10 years, depending on water use rate and quality. Screen inspections note any needed repairs. Lawns or hardscapes disguise underground systems from regular maintenance activities.

Filters and Screens

Filtration safeguards plumbing and irrigation emitters from particles unintentionally collected. **Pre-screening** uses coarse, tight-weave fabric over all inlets and vents, catching coarse debris. **Final Protection** installs before emitters comprising fine mesh or sediment sand media trapping smaller solids like silt, algae and insects plus particulate organics which may harbor pathogens.

Sizing Filters

Aperture sizes relate to intended uses, with drip emitters and misting nozzles necessitating superior screening over soaker hoses, for example. Guideline mesh/media pore sizes range between 150-to-600 microns, corresponding to typical impurities in harvested water and associated irrigation practices. Oversized filters allow bypassing, while undersized ones clog quickly.

Maintenance Practices

Unclogging periodically–from months to annually–involves backflushing or chemically treating fabric/sand media to remove built-up particles as water flows reverse through.chk Checking monthly for head loss alerts when pressure decreases, signaling needed cleaning. Replacing deteriorated materials averts breaks that release contaminants back into plumbing lines. Proper filtration forms the last line of defense for water quality post-storage.

Overflows and Outlets

Effective management of excess water is critical to the proper functioning of any rainwater harvesting system. This section will explore overflow and outlet design considerations in further detail.

Overflow Configuration

Proper overflow sizing is important to prevent tank overtopping during periods of heavy rain. Vertical pipes extending several inches above the tank rim are preferable to alternative designs to allow for easy cleaning and sediment removal. Elbow fittings on the outlet end help guide flow away from the tank surface to direct surges of water directly into the overflow drainage system.

Perforated vertical pipes placed inside the tank work well for concrete cistern construction. Plastic barrels can utilize exterior overflow tubing attached near the top rim. Fabricating an integral overflow component during tank fabrication eliminates the need for secondary attachment mechanisms.

Outlet Accessibility

Low-level drain outlets located near the base of storage vessels facilitate complete emptying for interior cleaning or repairs. Submersible ball valves or gate valves anchored securely to the tank base provide convenient access without the need for confined space entry. Drain tubing long enough to safely direct water flow away from personnel is also advisable from a safety and spill containment standpoint.

Access risers extending 12-18 inches above grade allow easy connection of garden hoses or suction lines. Watertight seals help prevent structural damage from surface water infiltration over time. Clear markings kept current on risers assist in the identification of functions over the lifespan of the installation.

Drainage Conveyance

Proper bedding material prevents damage to drainage lines from erosion, frost heaving, or root intrusion. Rigid pipe schedules 40 PVC, CPVC, or HDPE suit rainwater applications, while flexible corrugated piping works well for shallow runs. Slopes maintained between 0.5-2.0% transport flows depending on material size.

Cleanouts positioned at every change of direction or where slopes flatten out permit the flushing of sediment loads. Redirecting drainage away from building foundations using overflow swales or dry wells upholds structural integrity. Easements or licensed drainage outfalls become necessary for off-site discharge in certain jurisdictions.

Oversizing Considerations

Calculating overflow pipe diameters factors in maximum anticipated flow volumes from large storm rainfall depths. The 100-year or 500-year events often govern sizing to prevent surcharging and back flooding. Upsizing 25-50% beyond the minimum required diameters provides a margin of safety.

Control structures such as flow restrictors and flapped backflow preventers can regulate discharge velocities, which otherwise may cause erosion or flooding concerns if directed onto flat ground. Vegetated buffer strips also dissipate the energy of high-velocity overflow flows.

Overflow Release Points

Directing excess flows into pre-existing drainage features takes advantage of low-impact development techniques. Level spreaders and riprap energy dissipaters protect outlets from scour. Outlet protection, ranging from gabion baskets to loosely placed stones, prevents erosion at the point of release.

Level spreading away from slopes using shallow trenches filled with stone sets overflow water onto stable areas to recharge the landscape naturally. Signs posted near outlets caution against wading or swimming for safety during storms. Regular maintenance keeps structures free from blockages.

Monitoring and Maintenance

Overflow pipes require occasional visual inspection and cleaning to flush any accumulated sediment. Checking drainage swales and channels after significant rain events ensures proper conveyance of surplus volumes away from tanks and structures.

Documentation of flow observations during heavy rains aids in improving designs during subsequent renovations if needed. Periodic valve cycling exercises outlet mechanisms and identifies any repairs required to prolong output component performance for decades of service. Proper upkeep forms the long-term outlook of overflow system reliability.

CHAPTER 6
RAINWATER QUALITY AND PURIFICATION

The Roberts family was sitting under the clear night sky, enjoying fresh lemonade made from the crystal-clear rainwater collected in their 2500-gallon storage tank. Little did they know about the processes their collected rainwater went through to achieve this purity. Let's rewind to see what efforts lay behind the scenes to ensure their water was safe for consumption. This chapter delves into the importance of rainwater quality and various methods used for purification.

Common Contaminants

Rainwater purity is crucial for health, but its open collection means potential microbial and chemical contamination from various sources. Before using collected rain, understanding these risks is important. Contaminants pose differing risks based on various factors. This section delves deeper into common contaminants in greater detail.

Airborne Pollutants

Airborne pollutants settling on collection surfaces introduce a variety of chemical and particulate contamination into rainwater. Vehicular emissions contain nitrogen oxides, sulfur dioxide, carbon monoxide, heavy metals, and volatile organic compounds expelled through vehicle exhausts. Regular traffic near catchment areas exposes rooftops to these particulates. Improperly tuned vehicles or those using substandard fuels emit disproportionately higher levels of pollutants. Particulate matter of less than 2.5 microns can penetrate deep into the lungs, affecting respiratory and cardiac health on ingestion. Some metals like lead are toxic even at low doses, posing neurological risks. Polycyclic aromatic hydrocarbons in soot are carcinogenic with prolonged exposure. Stringent emissions testing and electric vehicles can help curb this source of contamination. Industrial regions face worse air quality issues depending on local industries and regulations. Power generation, oil refineries, smelters, and other factories releasing pollutants into the air worsen contamination risks. Old paint containing lead on surfaces also chips away, contributing to airborne heavy metals. While absolute elimination may not be feasible, selecting catchment sites carefully away from major emission points and vegetation filters can minimize air pollutant amounts in collected rainwater.

Bird, Animal, and Insect Droppings

Droppings from birds, bats, and other animals introduce pathogens, parasites, and contaminants into rainwater systems. Bird droppings may contain over 60 disease-causing organisms harmful to humans, such as Salmonella, E. coli, Cryptosporidium, and Chlamydia. Some birds, like pigeons, harbor drug-resistant strains, making infections difficult to treat. Even a single dropping is all it takes to contaminate the entire storage amount. Bats risk rabies virus exposure through any contact with saliva or droppings. Their guano may introduce fungal histoplasmosis inhaled as spores. Rodents carry a host of diseases transferred through urine and feces, including hantavirus, leptospirosis, and rat-bite fever. Their droppings support fungal growth, raising allergen and mycotoxin risks. Insect carcasses and cocoons left behind when hatching secrete pathogens within degrading tissues. Regular rooftop cleaning helps remove this bio-waste before rains flush it into the collection system. Installation of screens and seals limits access by dropping sources. Choosing smooth corrosion-resistant roofing materials discourages nesting and roosting, to begin with. Removal of trees offering perches and proximity to natural open areas helps control urban bird flock presence.

Leaf Debris and Plant Pollen

During rains, leaves and flowers washing off roofs carry pollen, molds, yeasts, bacteria, and leaf diseases into storage. Rotting leaves release tannins and lignin, imparting unpleasant tastes and odors to stored rainwater when degraded by microbes. Leaf fragments clog gutters, filters, and tanks, requiring more maintenance effort. Species like oak, pine, and sycamore shed pollen in huge quantities that deposit onto surfaces. Pollen grains transport fungal spores and allergens, triggering seasonal symptoms in sensitized persons. Certain molds thriving on moist, nutrient-rich leaf litter generate mycotoxins, posing chronic health issues with long-term exposure. Routine gutter cleaning before the onset of leaf drop season helps minimize the infusion of this organic matter into collected rainwater supplies. Deciduous trees shedding in fall-winter poses maximum contamination risk, and early leaf removal lowers that potential. Location factors like wind patterns and proximity to such vegetation also affect the extent of leaf wastes washed into the rainwater system with each rainfall event.

Mosses and Algae

In shaded, constantly damp areas prone to low flow velocities, mosses, and algae gain a foothold on surfaces, slowly spreading. Their growth accelerates with the availability of sunlight, moisture, convenience, nutrients, and no competition. Mosses like Sphagnum absorb 20 times their weight in water, trapping other particulates, which release decomposition. Some varieties accumulate heavy metals in tissues, rendering them hazardous when inhaled as fungal spores. Algal mats get shredded during rains, sloughing cells housing parasites, bacteria, endotoxins, and algal toxins released under stress. Species like Gloeotrichia produce geosmin, imparting an organic, muddy odor noticed in stored water, even at low levels, which lingers on objects washed or cooked with such supply. Decay of this plant growth also depletes oxygen levels in confinement, promoting the growth of anaerobes. Regular cleaning and removal of stagnant water helps curb moss-algae havens before they substantially contaminate rainwater quality. Modifications like sloped roofs and unobstructed gutter channels better facilitate self-cleaning during rains, too.

Chemicals From Roofing Materials

New asphalt, old lead paint, galvanized roofing, and even cement gradually leach certain compounds soluble in water when repeatedly wetted during rains. Fresh asphalt emits coal-tar pitch volatiles like benzene, toluene, and methyl benzenes hazardous to health. With aging, it releases polycyclic aromatic hydrocarbons, some of which cause cancer. Lead formerly added to paints for brightness gets absorbed into runoff over time, posing high poisoning risks, especially for children. Even low lead doses can lower IQ and cause behavioral issues. Leaching worsens under acidic conditions. Galvanized or zinc roofs corrode, eventually contaminating water with zinc particulate and soluble Zn^{2+} ions toxic in excess amounts. Cement also contains soluble oxides of calcium, silicon, aluminum and iron, besides heavy metals like nickel responsible for concrete cancer. Limiting the use of chemical-leaching materials, sealing potential seams, and frequently monitoring maintenance and replacement prevents the accumulative introduction of hazardous compounds into caught rainwater supplies.

Accidental Sewage Contamination

While rare, improper plumbing connections allowing sewage backflow during heavy rain or brimful conditions, cracked septic tanks discharging effluent onto catchment areas, and construction work in proximity breaching pipes and drains pose risks of rainwater contamination by untreated or partially treated human and animal waste. Sewage carries a whole host of enteric pathogens, including viruses like hepatitis A, various bacteria like Shigella and Salmonella, and pathogens including protozoa and helminths, causing diseases like cryptosporidiosis, giardiasis, and ascariasis. Many of these oocyst and cyst-forming organisms are highly chlorine-resistant, exacerbating treatment difficulties. Even a tiny accidental influx can expose consumers to infection risks, defeating the health benefits of collected rainwater usage. Careful site evaluation, redundant protection like backflow prevention valves installed pre-emptively, and periodic inspection checks help lower the chances of such an undesirable contamination occurrence, rendering stored water unsafe for consumption.

Filtration Methods

Filtration forms the cornerstone of any effective rainwater purification system by removing particulate contaminants that can harbor pathogens, sediments, debris, and cysts before disinfection treatments. This section explores common filtration techniques in greater depth.

Screens

Screens are the simplest yet most basic filtration level, utilizing a mesh of varying coarseness matched to the required retention rating. For roof runoff, a 1/8 inch or smaller mesh traps most foliage and grit, preventing entry to storage without compromising flow. Regular cleaning stays critical as screens clog fast, bypassing impurities. Stainless steel meshes withstand weather better than brass or mild steel but are prone to corrosion and deteriorating effectiveness over time. Self-cleaning screen backwash systems using valves exist for automatic, hands-free operation, but complexity adds costs. Due to large pore sizes, screens barely filter microscopic threats like bacteria, viruses, or protozoa. As pre-treatment alone, further filtration stages become indispensable for purified water goals. Location matters - screens placed before gutters collect maximum debris, extending service intervals compared to installation above tanks. However, predation reduces with elevated positions, keeping wildlife off easily accessed screens.

Sediment Filters

Deeper sediment cartridge designs featuring pleated synthetic media or wound fiber layers precisely engineered to specific micron ratings over 5μ effectively remove cysts, parasites, inorganic silt, and finer organic particles that slip through basic screens. Cotton, polypropylene, and melt-blown microfibers are common materials. With pore sizes down to 1μ, these depth filters consolidate the mechanical and physical processes of screening and sedimentation, achieving better clarity. But cartridges, too, require periodic washing or replacing spent media clogged with retained solids. Self-cleaning models backwashing with stored water automatically recharge usable life, saving manual labor. Sub-micron bentonite or perlite clay packets are inserted to provide additional adsorptive sedimentation, polishing wastewater to higher quality. Pilot studies showed 9-12 months of dependable field operation with monthly backwashing before requiring replacement.

Carbon Block Filters

Made from compressed powdered carbon, these dense black monolith filters chemically treat water through adsorption and absorption onto internal porous surfaces and microcavities. Activated by high heat, carbon blocks develop an extensive honeycomb structure with an immense cumulative surface area that effectively traps and holds a wide spectrum of soluble impurities below detectable levels. These include residual chlorine, organic molecules, odors, toxins, and tastes leached from materials maintaining freshness. With a service life spanning 6-12 months under normal usage, cartridge replacement remains simpler than carbon granular media requiring manual cleaning or replenishment. However, initial costs are higher than basic filters. Sediment pre-filtration prolongs carbon block life by lowering particulate loads and preventing premature clogging and channeling.

Membrane Filters

An advanced mechanical filtration method, membrane filters exploit semi-permeable polymeric films with uniform sub-micron pores to separate particles from fluids. These may be organic hollow fibers, ceramic or metal monoliths, and spiral-wound or plate-frame pressurized modules. Typical pore sizes range from 0.1 – 1 micron, catching even viruses as small as 27nm. Some specially developed MEMCOR hollow fibers reject 99.9999% of all microbial cysts, bacteria, viruses, and particulates down to 1nm, far surpassing USEPA standards. By size exclusion, they deliver absolutely clear, high-purity permeate while retaining suspended and colloidal solids. Automatic low-pressure membrane systems save space over drainable pressure vessels. However, membrane filters require skilled maintenance and frequent integrity testing or replacement of expiration-dated cartridges for continued efficacy. Pre-treatment is crucial to avoid premature fouling under silt loads.

Media Filters

Using naturally occurring filter media layered by effective size, multimedia filters provide an economical tertiary treatment polishing quality through gravitational sedimentation and finer mechanical straining. Best suited for larger commercial settings, a typical multimedia filter contains graded sand, anthracite coal and garnet aggregate in descending particle size sequence. As water percolates through randomly stacked media beds, particles bridging between granules get caught, producing clear permeation. Organic matter also develops biofilms that biologically decompose over weeks. Backwashing expels retained solids occasionally. While offering good service life, media filters require regular chlorine disinfection to control bacteria regrowth, establishing in moist filter beds if left idle for longer durations.

Purification Techniques: Boiling, UV, and Chlorination

Even after multiple filtrations pass, residual risks may persist, calling for additional microbial disinfection critical to ensuring collected rainwater safety. This section examines boiling, UV treatment, and chlorination in greater depth.

Boiling

As the oldest mode of thermal disinfection, boiling water offers a simple, inexpensive method to deactivate all waterborne pathogens by elevating its temperature beyond 99-100°C, killing microorganisms almost instantly. The process does not involve any chemical addition. Bringing water to a rolling boil accompanied by steam bubbles covering the entire surface guarantees thorough heating, allowing natural air-cooling thereafter. Three minutes at a full boil destroys over 99.999% of Giardia lamblia cysts, Cryptosporidium parvum oocysts, Salmonella typhi, and other common bacterial, viral, or parasitic pathogens. Boiling remains equally effective regardless of pH, turbidity level, or other water quality attributes, staying completely chemical-free. However, it is energy-intensive when large volumes require processing daily and boiling carries scalding risks, too, if not handled carefully. Stored boiled water also stands susceptible to recontamination without residual protection over time. Therefore, other options offering continuous disinfection gain popularity for main water supplies.

Ultraviolet (UV) Disinfection

UV rays generated by low-pressure lamps contain short wavelengths deadly to microorganisms by penetrating cell walls and disrupting their DNA, preventing replication. Low-wattage 254nm UV sterilizers installed at storage outlets or faucet-mounted achieve a logarithmic reduction of viruses, bacteria, and some protozoa. Studies show Giardia reduction surpassing 99.9999%, higher than the USEPA four-log virus requirement with adequate exposure. UV sterilization processes work instantaneously without adding chemical by-products delivered through a plug-and-protect mechanism at ambient temperatures. Amalgam UV lamps rated 9-18 months require simple annual replacements, while newer LED emitters may serve 5-6 years. UV efficacy relies upon low turbidity since particulates can shield pathogens by absorbing sufficient radiance. Pre-treatment becomes vital to sub-1NTU clarity, allowing for effective log kill calculation. Flow rate and exposure time management guarantee that the necessary UV dosage is delivered. Correct installation and periodic lamp verification using test strips validate continuous function.

Chlorination

Chlorine disinfectants like Sodium hypochlorite in liquid bleach form and calcium hypochlorite granules or gas apply oxidative destruction of viral envelopes, bacterial cell walls, and protozoal cysts/oocysts through breakpoint chlorination. A minimum chlorine residual maintains continuous protection against re-growth if any pathogens survive initial treatment. Standard measured chlorine dosing generates 3-5ppm free available chlorine for at least 30 minutes with undiminished contact time to achieve over 99.99% Giardia and virus inactivation. However, excess chlorination risks forming Trihalomethanes which are known to impact animal health with long-term exposure. Monitoring chlorine residual levels ensures adequacy without over-dosing. Dechlorination using Sodium thiosulfate neutralizes excess chlorine, preventing harmful exposure while dissipating residuals over time and requires boosting via intermittent rechlorination. Chlorination suits large continuous-use systems but necessitates proper handling, storage, and safety protocols for chlorine-based disinfectants.

Ozone Disinfection

Ozone gas (O3) produced from oxygen mediated by ultraviolet lamps or corona discharge introduces an equally powerful yet rapid disinfecting agent penetrating cell structures oxidatively. Despite the necessity of large ozonation equipment, it completely degrades into O2, leaving no organic by-products within minutes after treatment. A typical dosage achieving 99.99% inactivation of viruses, bacteria, and Giardia requires 1.5-2.0mg/L ozone contact for 15-20 minutes. Higher doses prove more effective against Cryptosporidium oocysts resistant to chlorine. However, sophisticated monitoring controls ozone concentrations precisely while minimizing the formation of toxic by-products from excess ozonation. As an unstable gas, residual protection becomes minimal, requiring subsequent disinfection via UV or chlorine before distribution for uses demanding extended microbial safety. Overall, higher capital costs and operational complexity relegate ozonation commonly for large water works applications.

Testing Water Quality

Regular water quality monitoring reinforces multiple treatment barriers' effectiveness by validating purification success and identifying any deviations early, allowing timely corrective actions. This ensures dependable portability for all intended uses.

Coliform Bacteria

Indicator microorganisms like total coliforms and specifically E. coli afford rapid, economical screening of treatment reliability and potential bacterial contamination given their ubiquitous environmental existence. IDEXX or Colilert-defined substrate technology field kits fluoresce or change color in 24 hours if coliforms are present. Laboratories conduct Membrane Filtration analysis detecting even a single coliform colony. While coliforms themselves may not cause illness, their presence signals treatment lapses, potentially allowing entry of pathogens harbored similarly. Testing monthly or bimonthly as required by regulations reassures safety. Observing absence validates consistent limits of detection disinfection standards. Any positive shows hygienic maintenance issues need resolution to forestall risks.

Turbidity directly relates to particle counts, obscuring clarity, and poses challenges to effective disinfection by shielding pathogens within flocs. Benchtop or portable instruments express turbidity in Nephelometric Turbidity Units (NTUs) by passing a beam through samples and quantifying light scattered by suspended materials. Drinking water standards mandate ≤1 NTU for systems employing disinfection. Testing weekly allows timely response to rising turbidity from precipitation events, pipe disturbances or treatment upsets. With each rainfall, quick results determine when pre-filtration requires readjusting for optimal clarity ahead of the purification stages. Maintaining awareness about monsoonal fluctuations safeguards microbiological safety.

pH

This scale between 0-14 measures hydrogen ion concentration indicating acidity or alkalinity affecting water chemistry parameters essential treatment and plumbing integrity. pH meters calibrated regularly against standard buffer solutions furnish reliable values. Since most processes function best between 6.5 and 8.5, monthly checks detect deviations endangering treatment chemical efficacy, equipment, or metallic distribution components through corrosion over the long term. Adjustments using sodium hydroxide or sulfuric acid quickly remedy unsuitable pH, permitting smooth operation.

Residual Disinfectants

Where chlorination forms part of the process, tests determine free chlorine concentration through colorimetric DPD methods using comparable field kits as laboratory procedures. Desired 0.2-2.0

mg/L levels guarantee continuing contact time protection when taking samples from throughout the premises. Tests run daily initially and then weekly to ensure no drop-offs occur, impairing safety before subsequent use. For ozone, indigo colorimetric analysis measures satisfactory residual levels upon leaving treatment to certify transit protection. Such monitoring validates ongoing disinfection maintained for user confidence in tap water microbial safety.

BOD5 and COD

Biochemical (BOD5) and chemical oxygen demand (COD) assays quantify organic pollution load-straining treatment processes. Here, dilute samples incubate for 5 days as aerobic bacteria metabolize dissolved organics under DO monitoring or get titrated with dichromate/sulfuric acid, indicating oxidation requirements. Testing monthly aids advance notification for necessary filtration upgrades when limits surpass 10mg/L (BOD5) and 60mg/L (COD) set for drinking water. Lower readings attest to effective pretreatment and purification sustaining good quality over the long term.

Regular water testing through prescribed standard methods forms the backbone of any prudent catchment water system quality assurance program. Beyond mandated regulatory conformities, it aids proactive decision-making, enhancing public health protection through waterborne illness prevention.

CHAPTER 7
DESIGNING YOUR RAINWATER HARVESTING SYSTEM

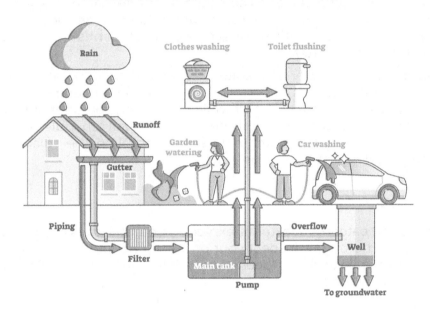

A solid foundation is crucial before erecting any structure. Likewise, properly designing your rainwater harvesting system is essential for establishing an efficient and durable setup. Just as an architect meticulously plans a building's blueprint, taking into account the location, purpose, and long-term usage, developing a customized design for your rainwater harvesting needs careful consideration. This chapter guides you through the key decisions to make and factors to keep in mind for designing a system tailored to your unique context and requirements.

Basics of System Design

To begin designing your rainwater harvesting system, it is important to identify the key factors that need consideration. Clearly outlining your objectives and assessing your site conditions allows for a customized plan tailored to suit your needs. While some homework is required upfront, taking the time for thorough pre-planning leads to an optimally functioning long-term solution. Let's explore the fundamental elements of system design.

Estimating Water Demands

Calculating your projected water requirements is essential for appropriately sizing the system's storage capacity and components. To estimate indoor demand, analyze your household's daily water usage from activities like cooking, drinking, laundry, bathing, and cleaning and determine the total liters needed. Factor in the number of residents and any planned changes. For longer-term planning, consider potential expansions like additional residents or conversion of a garage into a living area that could increase indoor needs over time.

Outdoor demands also require quantification separately. Calculate garden and lawn irrigation needs based on the types of plants, their watering schedules, and total areas. Compute volumes required for other outdoor applications like washing cars, paths, livestock watering, etc. Aggregating anticipated daily demands for individual purposes helps forecast the total harvestable water potentially required on average as well as peak days.

Assessing Catchment Areas

Identifying all potential catchment surfaces is an important early step. Carefully measure dimensions of roofing, patios, or other collection areas to compute their total catchable rainfall potential, which influences the harvesting capacity. For roofs, record and add up dimensions of separate sections if there are multiple levels or wings. Confirm surfaces are intact, debris-free, and correctly sloped to maximize redirected runoff into gutters.

Additionally, inspect materials to verify their suitability and safety for potable use catchment as per local regulations. Some roofing sheets contain toxic materials and need protection from leachates. Smooth surfaces promote faster, cleaner runoff while preventing contaminants from being washed in. Making notes of catchment areas, their attributes, and overhaul schedules assists long-term planning and maintenance timelines. With catchment sizing complete, their placement relative to the storage tanks can be considered at the design stage.

Identifying Optimal Locations

Given identified needs and catchment extents, exploring potential sites for the rainwater infrastructure becomes important. Thoroughly inspect your property, mapping out prime locations for tanks and tap points, factoring terrain and intended usage points. Your home's unique topography plays a key role in available layout options.

For a gravity-fed system, shortlist sites immediately downhill of catchment areas and above points of use to enable unassisted flow. Consider necessary separation distances between components for maintenance access as per regulations, too. Alternatively, if pumping is preferable, select spots allowing easy equipment installation and future servicing while remaining accessible. Multiple smaller decentralized tanks may perform better than one large central unit, giving versatility.

Scrutinizing and pre-testing locations streamlines siting suitability before fixation, avoiding later expensive reworks. Note intended coverings, fencings, and environments of selected areas for future-proofing. With preliminary component positions identified, draft designs can visualize potential configurations.

Sketching Preliminary Designs

To this point, various steps have helped gather crucial information regarding needs, resources, and constraints. Now, it is time to synthesize learnings through rough schematic designs showcasing alternative layout options. Simple hand-drawn sketches suffice at this phase.

Accurately locate and proportion catchment areas, tanks, and tap points, factoring terrain rises and setbacks. Draft pipe alignments between components and note preliminary material selections. Include relevant constructional aspects like cover slabs or fencing as relevant.

Present designs to professionals or experienced rainwater catchment system owners for review and critical feedback. Refine or rework concepts based on received insights before finalizing the design approach acceptable to your requirements and site. Outline broad cost estimates of prospective configurations as another decision metric.

Iterations on paper at the concept stage save huge costs and effort of ripping out and replacing real physical infrastructure later due to flaws or issues missed on paper. The process delivers well-reasoned, implementable designs ready for the next stages of permitting and construction.

Gravity vs. Pump Systems

Two principal approaches can transport harvested rainwater - utilizing gravity through downhill pipelines or mechanically lifting it using electric-powered pumps. Each method presents advantages depending on your terrain and needs. Weighing their technical merits assists in the selection of the optimally suited system type.

Pros and Cons of Gravity Systems

Gravity flow conditioned by natural elevation differences provides trouble-free operation without electric running costs. Proper height differentials enable the effortless driving of water to its final destinations from catchment surfaces via storage tanks and then to tap points.

However, the terrain must perfectly comply through all phases for gravity's dependability. Even minor undulations can disrupt downhill motion requiring lifting, defeating its purpose. Correct falls need to be established right from the collection point to the highest usage area.

Furthermore, fixing leakages becomes difficult with non-removable installations underneath the ground. Access for maintenance tank cleaning also poses a challenge given non-portable positioning. But with apt topography, gravity remains the most convenient option.

Pros and Cons of Pump Systems

Mechanical pumping grants unmatched placement flexibility for tanks and piping interfaces independent of the land's levels and slopes. Components can be located anywhere as long as an electricity supply connects the system. Mobile pumps facilitate shifting arrangements as desired.

But power dependencies mean reliance on uninterrupted power sources. Blackouts disrupt operation until restoration, compromising reliability for critical needs. Regular maintenance of moving pump parts like lubricating and sealing involves additional effort compared to gravity's hands-off flow.

Electric running costs also accumulate over the decades, outweighing gravity's cost-free usage if sized adequately. Water leakage from pressurized pipes under pressure demands a faster response to prevent wastage and further issues.

Choosing the Right Approach

For locales with naturally cooperating slopes, the gravity-driven approach perfectly delivers water hassle-free. However, undulating terrain restricts gravity siting, necessitating pumps to leverage inconsistent falls. Moreover, certain catchment or usage points may stay inevitably elevated regardless of location, compelling pumping.

Weighing the site-suitability, functionality, long-term economics, and self-reliability of both aids in selecting the most fitting system. With gravity highly terrain-dependent, its feasibility relies on a thorough evaluation of property topography, ideally surveyed professionally if complex. A hybrid design profiting both means requires careful balance.

Further understanding operational intricacies and annualizing lifetime expenditure assists the decision. Cost-benefit analyses, seeking expert consults, and pre-testing prototypes help finalize the prime technology basis before costly implementation commitments are made. The right choice eases future management and adaptations.

Maintaining Pumping Systems

To extract the pumping system's useful lifespan, disciplined aftercare remains critical. Clean debris-clogged intake filters regularly per maintenance schedules. Monitor lubricant levels and seal conditions, replacing as prescribed or when leakages appear. Inspecting pipelines helps detect unwelcome air pockets or ingress points.

Damages due to normal wear-and-tear, like mechanical failures, merit quick repairs to restore normal water transmission. Portable backup pumps provide uninterrupted water supply if primary units unexpectedly fail. Solar or generator power backups safeguard against electricity blackouts disrupting critical needs.

Energy consumption profiling assists optimization through upgrades like variable speed controls or switching to renewable sources like solar or hydropower. Recording running hours aids pump replacements before breakdowns. Periodically desilting suction lines prevents clogging. Adopting commonsense preventive steps supports long and dependable service.

Above-Ground vs. Underground Storage

Once harvested, rainwater demands safe containment until forthcoming usage. Tanks buried underground or constructed above ground each have merits reliant on conditions. Informed selection directs the most fitting technique based on functional and site-specific factors.

Pros of Above-Ground Tanks

Surface construction eases invaluable access for thorough cleaning, inspection, and repairs when demand arises. Unearthing tanks lay bare all internal segments for examination and servicing without confinement stresses.

Modular additional compartments simply stack over others, extending capacity without reworking groundwork. Installation crew and equipment requirements were also reduced versus excavations. Visual checks detect algal growths or leakages without special equipment.

Above-ground siting further allows repositioning tanks effortlessly by disassembling and shifting constituents piece-by-piece. Expansions fit contexts changing over decades. No diffusion of the structure ensues within soil over a lifetime, either.

Cons of Above-Ground Tanks

However, exposed containers may pose an unattractive site eyesore displeasing certain settings such as aesthetically-conscious neighborhoods. Risks exist from tampering, accidental damage, or health issues if attracting insects and bird nesting.

Without insulation, extensive surface area transmits higher heat-inducing evaporation losses than encapsulated below-ground tanks. Tall constructed tanks require reinforced stabilization to withstand wind forces and earthquake risks compared to buried counterparts. Safety enclosures increase costs.

Pros of Underground Tanks

Underground installations vanish tanks from plain view, blending discreetly into most landscapes. Apart from initially displaced soil, the subterranean structure stays essentially invisible for a long time.

Burial naturally insulates stored water, minimizing surface area contacting sun and air to reduce evaporation to negligible levels. Protection from tampering, damage by falling objects, and microbial contamination arise inherently. Structural strength demands reduce significantly without wind loads.

Cons of Underground Tanks

Major excavations for large-volume buried tanks escalate initial installation expenses, requiring heavy construction equipment. Refilling voids afterward constitutes a whole separate operation.

Difficult access constraints cleaning and maintenance, increasing spending on repairs or tank overhauls if issues arise internally. Detecting leaks or intrusions becomes challenging without visual confirmation abilities.

Structural integrity must factor soil and hydrostatic pressures not affecting above-ground tanks. Tank material needs to resist corrosive soils and underground microbes over many rainy seasons. Anchoring resists floating risks if water-logged.

Scalability and Future-Proofing Your System

When you design a rainwater system, you want it to work well for a long time. Making it scalable means you can easily expand it as needed. This way, you don't have to replace the whole system when you want to make it bigger.

Scalable Options for Catchment

For collecting rainwater, you can add more surfaces like roofs to catch more water. You can also use permeable paving for patios and driveways to get more water into the system. Other options include rain gardens, swales, and solar stills to collect water efficiently. Planning ahead and making sure everything fits together makes it easier to add new parts without big changes.

Scalable Options for Storage

When it comes to storing water, using modular tanks that can connect to each other lets you increase the storage step by step. Some tanks have double walls and sacrificial cores, which makes it possible to use them for a long time. Systems that transfer water between tanks help keep everything in balance, and extra pumps can move water to the next set of tanks. Access chambers between tanks make it easy to add more tanks without causing problems. Doing all this saves you time and money in the future.

Scalable Distribution Systems

When setting up water distribution systems, it's better not to make them too big at first. Large networks and access points might distribute more water in the future without needing a lot of changes. The layout should be spread out like a mesh, with valves in different places, so it's easy to connect more parts without spending too much.

Control valves are important for managing pressure and keeping the system working well, as more water is needed over time. Recycling water helps save resources, especially as ways to clean water get better. When planning the roads for maintenance, think about the vehicles that will be needed for bigger systems in the future.

Future-Proofing Finances

To make sure you can afford water in the future, it's important to predict how much water will be needed. Projecting costs and checking bills against budgets helps keep monthly payments low. Having extra savings for unexpected costs is also a good idea.

Spreading out the cost of big investments over time can reduce the overall interest you have to pay. Some options, like trading rainwater credits, encourage communities to work together. Making sure you have financial support that matches your plans is important. Planning ahead wisely helps sustain long-term goals.

Being flexible is crucial to making changes that make sense economically. This helps protect people's jobs and the environment during expected changes. Systems that work well with nature can change and grow smoothly over time.

CHAPTER 8
STEP-BY-STEP PROJECTS FOR THE HOMESTEADER

Embarking on the journey of harvesting rainwater for your homestead is a thrilling venture. In this chapter, we delve into the practical side of rainwater harvesting, offering step-by-step guidance on various projects that cater to enthusiasts of all expertise levels. From the simplicity of a basic rain barrel collection to the intricacies of an underground rainwater storage system, each project is crafted with the homesteader in mind. Join us as we narrate a day in the life of a DIY enthusiast, brimming with excitement and passion for constructing rainwater harvesting projects that not only contribute to self-sufficiency but also champion sustainable living.

Simple Rain Barrel Collection: A Deep Dive into the Basics

Let's follow Alex, our DIY enthusiast, as they embark on setting up a basic rain barrel collection system. Picture a clear morning after a night of rain, the air crisp with the scent of wet soil. Alex, fueled by excitement, begins their day armed with the essential tools and a vision of harnessing rainwater for their homestead.

Step-by-Step Guide

1. **Gathering Materials:** The first crucial step in setting up a rain barrel collection system is acquiring the necessary materials. Here's what you'll need:

- **Rain Barrel:** At the heart of the system, the rain barrel should have a secure lid to prevent debris and mosquitoes from entering.
- **Downspout Diverter Kit:** This kit facilitates the redirection of rainwater from the downspout into the barrel.
- **Sealant:** A waterproof sealant ensures a tight connection between the diverter and the downspout.
- **Level:** To ensure the rain barrel sits evenly and stably.
- **Drill:** Necessary for creating openings in the barrel for the diverter and overflow connections.

2. **Choosing a Location:** Selecting the right location for your rain barrel is critical for its efficiency. Consider the following when choosing a spot:
   - **Proximity to Downspout:** Place the rain barrel near a downspout with easy access to a large roof area for maximum water collection.
   - **Level Ground:** Ensure the ground is level to prevent the barrel from tipping over.
   - **Accessibility:** Choose a location that allows easy access for regular maintenance and use.

3. **Installing the Downspout Diverter:** The downspout diverter is an essential component that redirects water into the rain barrel. Follow these steps for proper installation:
   - **Attach the Diverter Kit:** Follow the manufacturer's instructions to attach the diverter to the downspout.
   - **Secure Connection with Sealant:** Use a waterproof sealant to secure the connection between the diverter and the downspout, preventing leaks.

4. **Placing the Rain Barrel:** Now that the diverter is in place, it's time to position the rain barrel beneath it:
   - **Positioning:** Place the rain barrel directly beneath the downspout, ensuring a seamless connection with the diverter.
   - **Using a Level:** Employ a level to ensure the barrel sits evenly on the ground.

5. **Connecting Overflow:** To prevent water from overflowing and potentially causing damage, it's crucial to install an overflow system:
   - **Overflow Valve**: If your rain barrel is equipped with an overflow valve, connect it to redirect excess water away from the foundation.
   - **Proper Connection:** Ensure a secure connection to prevent leaks.

6. **Securing the System:** A stable rain barrel is essential for its functionality and safety:
   - **Fastening:** Secure the rain barrel in place to prevent accidental tipping or displacement.
   - **Stability Check:** Double-check the level and stability of the system.

7. **Regular Maintenance:** To ensure the longevity and effectiveness of your rain barrel collection system, regular maintenance is key:
   - **Cleaning:** Periodically clean the barrel to remove debris and prevent clogs.
   - **Leak Check:** Regularly inspect the system for leaks and reseal connections if necessary.

Advanced Rainwater Collection with First Flush

Rainwater harvesting, in its advanced form, involves more sophisticated systems to ensure the collected water is of the highest quality. The incorporation of a first flush mechanism is a crucial step in this process, as it helps divert the initial runoff, which may contain contaminants, away from the main storage. In this detailed guide, we will walk you through each step of setting up an advanced rainwater collection system with a first flush, providing you with the knowledge to enhance the quality and efficiency of your rainwater harvesting efforts.

Step 1: Materials Needed

Before diving into the installation process, it's essential to gather all the necessary materials and tools. For an advanced rainwater collection system with a first flush, you'll need:

- Large capacity rain barrel
- First flush diverter kit
- Additional downspout extensions
- PVC pipes
- Hose clamps
- Saw

Ensure that all components are of high quality and suitable for the size of your harvesting system.

Step 2: Planning the System

A successful rainwater harvesting system begins with careful planning. Start by mapping out the route from the downspout to the rain barrel. Consider the layout of your property, the location of the downspout, and any potential obstacles or structures that might affect the installation. This planning phase is crucial for ensuring the system's efficiency and longevity.

Step 3: Installing the First Flush Diverter

The first flush diverter is a key component of advanced rainwater harvesting. Follow these steps to install it properly:

- Attach the diverter at the top of the downspout according to the manufacturer's instructions.
- Ensure a secure connection by using sealant or appropriate fasteners.
- Connect additional downspout extensions to direct water to the first flush system.

The first flush diverter works by capturing and diverting the initial portion of the rainfall, which may contain debris and pollutants, away from the main storage unit.

Step 4: Connecting to the Rain Barrel

Now, let's connect the first flush system to the rain barrel. This involves the use of PVC pipes and hose clamps to guide the filtered water from the first flush mechanism into the rain barrel. Follow these steps:

- Cut the PVC pipes to the required length using a saw.
- Connect the pipes to the first flush diverter and secure them tightly with hose clamps.
- Guide the PVC pipes to the rain barrel, ensuring a secure and watertight connection.

This step is crucial for channeling the filtered water into the main storage unit while keeping the initial runoff with contaminants away.

Step 5: Setting Up the First Flush Mechanism

The first flush mechanism is designed to divert the first portion of rainwater away from the rain barrel, preventing contaminants from entering the storage system. Here's how to set it up:

- Install a mechanism within the first flush system that allows the initial water to flow separately.
- This could be a simple valve or a more sophisticated design, depending on the diverter kit.
- Ensure that the mechanism is functioning correctly and diverts the initial runoff away from the rain barrel.

Properly setting up the first flush mechanism is vital for maintaining the quality of the harvested rainwater.

Step 6: Securing the System

To ensure the longevity and stability of your advanced rainwater collection system, take the following precautions:

- Double-check all connections to ensure they are secure and leak-free.
- Fasten any loose components, such as PVC pipes or hose clamps.
- Ensure the first flush diverter is securely attached to the downspout and the additional extensions.

A well-secured system not only functions better but also minimizes the risk of damage or malfunctions over time.

Step 7: Testing the System

Before considering the installation complete, it's crucial to test the system to ensure it functions as intended. Here's how to conduct a thorough test:

- Simulate rainfall by using a hose or waiting for the next rain event.
- Observe the first flush mechanism in action, making sure it diverts the initial runoff away from the rain barrel.
- Check for any leaks, especially at connection points, and address them promptly.
- Adjust the system if necessary for optimal performance.

Testing the system under simulated rainfall conditions allows you to identify and address any issues before relying on it during actual weather events.

Underground Rainwater Storage System: Harnessing Earth's Bounty

Step-by-Step Guide

✓ Materials and Tools:

Before delving into the excavation and installation process, gather the essential materials and tools required for this ambitious project:

- **Large underground tank:** Select a tank with sufficient capacity to meet your rainwater storage needs. Tanks made of durable materials like polyethylene or concrete are ideal.
- **Excavation equipment:** Depending on the scale of the project, you may need a backhoe, mini-excavator, or even manual tools for smaller installations.
- **Geotextile fabric:** This fabric acts as a barrier, preventing soil particles from infiltrating the tank and compromising water quality.
- **Gravel:** An important component for drainage, gravel facilitates the movement of water around the tank, preventing stagnation and potential structural issues.
- **PVC pipes and connectors:** These are vital for guiding rainwater from the downspout to the underground tank.
- **Concrete or bricks**: These materials provide a stable foundation for the tank and contribute to the overall structural integrity of the system.

✓ Planning the Excavation:

The success of an underground rainwater storage system hinges on meticulous planning. Before breaking ground, consider the following:

- **Mark the area for excavation:** Using stakes and string, outline the dimensions of the pit. Be sure to adhere to local building codes and regulations.
- **Contact utility services:** Before excavation begins, contact utility services to mark the location of underground utilities. Safety is paramount, and this step prevents potential accidents during the digging process.
- **Determine the depth and size:** The depth and size of the pit should align with the capacity of the chosen tank and the expected rainfall in your region. A depth of at least 6 feet is recommended to provide adequate storage capacity.

✓ Digging the Pit:

Once the planning is complete, it's time to break ground:

- **Using excavation equipment:** If using heavy machinery, carefully operate the equipment, ensuring precision in digging. Gradually remove the soil, keeping the sides of the pit as vertical as possible.
- **Manual excavation:** For smaller installations, manual labor may be necessary. Dig the pit with shovels, ensuring uniformity in depth and dimensions.
- **Creating a spoil pile:** As you excavate, create a spoil pile to store the extracted soil. This will facilitate the backfilling process later.

✓ Preparing the Base:

The success of an underground rainwater storage system depends on a stable and well-drained base:

- **Lining the pit with geotextile fabric:** Spread the geotextile fabric along the base of the pit. This acts as a barrier, preventing soil infiltration into the tank.
- **Adding a layer of gravel:** Place a layer of gravel on top of the geotextile fabric. This layer enhances drainage and provides additional stability to the tank.

✓ Installing the Tank:

The moment has arrived to introduce the heart of the system - the underground tank:

- **Lowering the tank into the pit:** Carefully lower the tank into the prepared pit. It is crucial to ensure that the tank sits level and stable to prevent structural issues.
- **Connecting the downspout:** Use PVC pipes to guide rainwater from the downspout to the underground tank. Install a filter in the line to prevent debris from entering the tank.

✓ Covering and Securing:

To protect the underground tank and maintain its stability, the following steps are crucial:

- **Covering the tank:** Once the tank is in place, cover it with additional layers of geotextile fabric. This adds an extra layer of protection against soil infiltration.
- **Building a protective barrier:** Surround the tank with concrete blocks or bricks, leaving access points for future maintenance. This protective barrier shields the tank from external pressure and provides stability.

✓ **Testing the System:**

Before considering the project complete, it's imperative to run a series of tests:

- **Checking for leaks**: Simulate rain by introducing water into the system and carefully inspect for any leaks. Address any issues promptly to avoid compromising water quality.
- **Ensuring proper water flow:** Confirm that rainwater flows seamlessly from the downspout through the PVC pipes and into the underground tank. Adjust the slope of the pipes if necessary.

Integrating Rainwater into Home Plumbing: A Comprehensive Guide

Assessing Existing Plumbing

✓ **Purpose:**

Before delving into the integration process, it's crucial to understand the layout of your existing plumbing system. This step helps identify the main water line entering your house and plan the integration points for rainwater.

✓ **Step-by-Step Guide:**

1. **Locate the Main Water Line:**
   - Identify the point where the main water line enters your home.
   - This is typically near the water meter or where the municipal water supply connects to your property.
2. **Sketch a Plumbing Layout:**
   - Create a schematic diagram of your existing plumbing system.
   - Highlight key fixtures such as toilets, sinks, washing machines, and outdoor faucets.
3. **Identify Integration Points:**
   - Determine which fixtures are suitable for rainwater integration.
   - Prioritize non-potable water applications like toilets, washing machines, and outdoor use.

Installing the Filtration System

✓ **Purpose:**

A robust filtration system is crucial to ensure the quality of harvested rainwater meets acceptable standards for household use. This step involves installing a filtration unit to remove impurities and debris.

✓ **Step-by-Step Guide:**

1. **Select a Suitable Filtration System:**
   - Choose a filtration system designed for rainwater harvesting.
   - Consider factors such as filtration capacity and ease of maintenance.
2. **Identify Integration Point:**

- Install the filtration unit close to the point where rainwater enters your plumbing system.
- This is typically after the rainwater storage system and before the point of use.

3. **Connect to Main Water Line:**
   - Cut into the main water line at the identified integration point.
   - Install the filtration unit in line with the existing plumbing.

4. **Install Check Valves:**
   - Place check valves before and after the filtration unit to prevent backflow.
   - This ensures that rainwater and municipal water do not mix.

5. **Seal Connections:**
   - Use appropriate sealants to ensure watertight connections.
   - Check for leaks and tighten connections as needed.

Integrating the Pressure Tank

✓ Purpose:

A pressure tank helps regulate water pressure within your plumbing system, ensuring a consistent flow of water to your fixtures. This step involves integrating a pressure tank into the rainwater harvesting system.

✓ Step-by-Step Guide:

1. **Select a Pressure Tank:**
   - Choose a pressure tank that suits the size of your household and water usage needs.
   - Ensure it is designed for use with rainwater.

2. **Identify Integration Point:**
   - Determine the optimal location to install the pressure tank.
   - This is usually near the point where rainwater enters your plumbing system.

3. **Connect to Filtration System:**
   - Install the pressure tank downstream from the filtration unit.
   - Use PVC pipes and connectors to establish a connection.

4. **Install a Water Pump:**
   - Connect a water pump to the pressure tank.
   - Ensure the pump is designed for use with rainwater and can handle the required flow rate.

5. **Establish Electrical Connections:**
   - If using an electric pump, connect it to a power source.
   - Ensure compliance with local electrical codes.

6. **Adjust Pressure Settings:**
   - Set the pressure tank to maintain adequate water pressure throughout your plumbing system.
   - Consult the tank's manual for specific adjustment instructions.

7. **Test the System:**
   - Turn on fixtures supplied by rainwater and observe the system's performance.
   - Adjust pressure settings as needed.

Connecting Rainwater Source

✓ Purpose:

This step involves guiding rainwater from the storage system to the filtration unit, ensuring a secure and watertight connection.

✓ Step-by-Step Guide:

1. **Select Suitable PVC Pipes:**
   - Choose PVC pipes of appropriate diameter for optimal water flow.
   - Ensure the pipes are rated for outdoor and underground use.
2. **Identify Connection Points:**
   - Determine where rainwater will enter the plumbing system.
   - Cut into the downspout or use a dedicated pipe connected to the rainwater storage.
3. **Install a Filter:**
   - Place a filter at the entry point to prevent debris from reaching the plumbing system.
   - Clean or replace the filter regularly to maintain efficiency.
4. **Connect Pipes to Filtration Unit:**
   - Use PVC pipes and connectors to establish a connection between the rainwater source and the filtration unit.
   - Ensure a gradual downward slope to facilitate water flow.
5. **Seal Connections:**
   - Apply sealant to all connections to prevent leaks.
   - Regularly check and reseal connections as part of routine maintenance.

Routing to Household Fixtures

✓ Purpose:

This step involves establishing connections to direct rainwater to specific fixtures within your home, such as toilets, washing machines, and outdoor faucets.

✓ Step-by-Step Guide:

1. **Identify Fixtures for Integration:**
   - Select fixtures that are suitable for rainwater use.
   - Prioritize non-potable water applications to maximize efficiency.
2. **Cut into Existing Plumbing:**
   - Cut into the existing plumbing lines that supply the identified fixtures.
   - Ensure precise cuts to avoid unnecessary disruptions.
3. **Install Additional Check Valves:**
   - Place check valves on the rainwater side of the plumbing lines.
   - Prevent backflow and maintain separation between rainwater and municipal water.
4. **Connect Rainwater Lines:**

- Use PVC pipes to establish connections between the rainwater supply and the fixtures.
- Install shut-off valves for each connection point.

5. **Adjust Flow Rates:**
   - Fine-tune the flow rates for each fixture to optimize water usage.
   - Ensure a balance between rainwater and municipal water supply.
6. **Test Each Fixture:**
   - Turn on each fixture supplied by rainwater individually.
   - Check for consistent water flow and address any issues promptly.

Creating a Rain Garden: A Flourishing Oasis of Sustainability

Selecting a Location

Choosing the right location for your rain garden is paramount to its success. It's a choreography between aesthetics and functionality. Here's how you can find the perfect spot:

a. Assess the Landscape:
- Survey your yard for low-lying areas where rainwater naturally accumulates.
- Avoid places prone to flooding, as the goal is to capture rainwater, not exacerbate drainage issues.

b. Consider Sunlight Exposure:
- Analyze the sunlight exposure of potential locations throughout the day.
- Aim for a spot that receives a mix of sun and shade, catering to a variety of plants.

c. Evaluate Soil Composition:
- Test the soil in the chosen area for drainage capabilities and nutrient content.
- The ideal rain garden soil is a well-draining mix of sand, silt, and organic matter.

Planning the Garden Layout

Once you've identified the ideal location, it's time to plan the layout of your rain garden. This involves a creative process that combines functionality with aesthetic appeal.

a. Sketching the Design:
- Consider the overall size and shape of the rain garden.
- Sketch a design that includes curves and contours to maximize water retention.

b. Selecting Plant Locations:
- Research native plants that thrive in rain gardens.
- Arrange the plants according to their water needs, placing those with higher water tolerance in lower-lying areas.

c. Ensuring Proper Drainage:
- Design the rain garden to have a gentle slope to encourage water flow.
- Create mini-basins or depressions to trap and hold water during rain events.

Excavating the Area

With your design in hand, it's time to bring it to life by excavating the designated area for your rain garden.

a. Marking the Boundaries:
- Use marking paint or stakes to outline the boundaries of the rain garden.
- Ensure the design aligns with the natural flow of rainwater.

b. Digging the Basin:
- Begin excavating the marked area, aiming for a depth of around 4 to 8 inches.
- Gradually slope the sides towards the center for efficient water retention.

c. Creating Berms:
- Use the excavated soil to create small berms or mounds around the edges of the rain garden.
- Berms prevent water from flowing out of the designated area.

Adding Mulch and Compost

With the basin in place, it's time to enrich the soil and create an environment conducive to plant growth.

a. Spreading Organic Mulch:
- Cover the basin with a layer of organic mulch, such as wood chips or shredded bark.
- Mulch helps retain moisture, suppresses weeds, and regulates soil temperature.

b. Incorporating Compost:
- Mix compost into the soil within the rain garden basin.
- Compost enhances fertility and provides essential nutrients for plant growth.

c. Mulching Again:
- After planting, apply a fresh layer of mulch around the base of each plant.
- Mulch helps maintain soil moisture and acts as a protective barrier.

Choosing Rain Garden Plants

Selecting the right plants for your rain garden is crucial for its overall health and success. Native plants are particularly well-suited to your local ecosystem.

a. Researching Native Species:
- Investigate native plants that thrive in the soil and climate of your region.
- Choose a mix of flowers, grasses, and shrubs for biodiversity.

b. Considering Water Tolerance:
- Group plants with similar water requirements together.
- Place those with higher water tolerance in the lower-lying areas of the rain garden.

c. Ensuring Seasonal Interest:
- Select plants that provide visual interest throughout different seasons.
- This ensures your rain garden remains appealing year-round.

Planting and Watering

Now that you've selected your plants, it's time to bring your rain garden to life by carefully planting and establishing a watering routine.

a. Planting Technique:

- Dig holes for each plant, ensuring they are at the same depth as they were in their nursery pots.
- Space plants, according to their mature size, avoid overcrowding.

b. Watering After Planting:
- Water the newly planted rain garden thoroughly.
- Maintain a consistent watering schedule during the first few weeks to help plants establish roots.

Mulching Again

Mulching isn't a one-time affair. Regular mulching is essential to the ongoing health of your rain garden.

a. Mulching After Planting:
- Apply a fresh layer of mulch around the base of each plant.
- Leave a small gap around the stem to prevent rot.

b. Periodic Mulch Renewal:
- Renew the mulch layer periodically to maintain its effectiveness.
- Mulching helps conserve soil moisture and suppresses weed growth.

Rooftop Garden Integration: A Verdant Oasis Above

Assessing Rooftop Structural Integrity

Detailed Steps:
- **Survey the Rooftop:**
  - Begin by conducting a thorough examination of the rooftop's structural condition. Look for signs of wear, weakness, or damage.
  - Pay attention to load-bearing capacity and ensure it can support the weight of a garden. If uncertain, consult with a structural engineer to assess the feasibility.
- **Check Local Regulations:**
  - Research and adhere to any local building codes or regulations regarding rooftop gardens. Some areas may have specific requirements to ensure safety and compliance.
- **Evaluate Accessibility:**
  - Consider how materials and plants will be transported to the rooftop. Ensure there's a safe and convenient way to access the space.

Creating Container Beds

Detailed Steps:
- **Select Suitable Containers:**
  - Choose containers that are lightweight yet sturdy, as they need to withstand environmental conditions while not adding excessive weight to the rooftop.
  - Options include plastic, fiberglass, or fabric pots. Ensure they have proper drainage holes.
- **Prepare the Soil:**
  - Use high-quality, well-draining soil to fill the containers. A mix of potting soil, perlite, and compost works well for container gardening.

- Consider adding slow-release organic fertilizers to provide nutrients over time.
- **Arrange Containers Strategically:**
  - Plan the layout of the containers to maximize sunlight exposure for the plants.
  - Place larger containers on the periphery to create a visual frame and smaller containers toward the center for variety.

Installing Drip Irrigation System

Detailed Steps:
- **Gather Drip Irrigation Components:**
  - Acquire a drip irrigation kit or individual components such as tubing, emitters, connectors, and a timer.
  - Ensure the system is suitable for container gardening and can be connected to the rainwater storage.
- **Lay Out the Tubing:**
  - Unroll the tubing along the rows of containers, making sure it reaches each plant.
  - Cut the tubing to the desired lengths, allowing for some slack for adjustments.
- **Install Emitters:**
  - Attach drip emitters to the tubing, placing them near the base of each plant.
  - Adjust the flow rate on each emitter to match the water needs of different plants.
- **Connect to Rainwater Storage:**
  - Use PVC pipes or flexible hoses to connect the drip irrigation system to the rainwater storage system.
  - Install a filter to prevent debris from clogging the emitters.

Selecting Plants

Detailed Steps:
- **Consider Container-Friendly Plants:**
  - Choose plants that are well-suited for container gardening. This includes vegetables, herbs, and flowers that don't have extensive root systems.
  - Research the sunlight and water requirements of each plant.
- **Create a Planting Plan:**
  - Sketch a layout for the rooftop garden, taking into account the size and growth habits of chosen plants.
  - Group plants with similar water and sunlight needs together.
- **Choose Local Varieties:**
  - Opt for plants native to the region or well-adapted to the local climate. This ensures they thrive in the rooftop environment.

Arranging the Garden

Detailed Steps:
- **Consider Aesthetic Layout:**
  - Arrange the containers in a visually appealing way, considering colors, textures, and heights.
  - Use taller plants as focal points and shorter plants as fillers.

- **Ensure Adequate Spacing:**
  - Avoid overcrowding the containers to allow each plant ample space for growth.
  - Consider the mature size of each plant when determining spacing.
- **Create Paths or Access Areas:**
  - Designate paths or clear areas for easy navigation around the rooftop garden.
  - Ensure there's enough space to comfortably tend to plants and enjoy the garden.

Connecting to Rainwater Storage

Detailed Steps:
- **Set Up PVC Piping:**
  - Use PVC pipes to guide rainwater from the storage system to the rooftop.
  - Consider the slope for gravity-assisted flow and secure the pipes to prevent movement.
- **Install a Filter System:**
  - Incorporate a filter system to ensure that debris and contaminants are removed from the rainwater before reaching the garden.
  - Regularly clean and maintain the filter for optimal performance.
- **Attach a Diverter (Optional):**
  - Include a diverter in the system to redirect excess rainwater when the containers are full.
  - This prevents overwatering and potential damage to plants.

Mulching and Fertilizing

Detailed Steps:
- **Apply Mulch:**
  - Cover the surface of the container beds with a layer of organic mulch.
  - Mulch helps retain moisture, suppress weeds, and regulate soil temperature.
- **Use Organic Fertilizers:**
  - Incorporate organic fertilizers into the soil or use liquid organic fertilizers in the irrigation system.
  - Follow recommended application rates to provide essential nutrients.

Regular Maintenance

- Regularly inspect plants for signs of pests, diseases, or nutrient deficiencies.
- Promptly address any issues to maintain a thriving rooftop garden.
- Adjust the drip irrigation schedule based on weather conditions and seasonal changes.
- Be mindful of rainfall and reduce watering during wet periods.
- Prune plants as needed to control growth and encourage bushier, healthier growth.
- Deadhead flowers to promote continuous blooming.
- Harvest vegetables and herbs when they reach maturity.
- Take time to relax and enjoy the beauty of the rooftop garden.

CHAPTER 9
USING COLLECTED RAINWATER EFFECTIVELY

Rainwater harvesting is an increasingly important practice for households aiming to be self-reliant and environmentally conscious. While the collection of this precious resource is a crucial first step, its effective utilization is equally important to maximize benefits. This chapter delves into diverse ways to utilize harvested rainwater around a homestead through garden and agricultural irrigation, indoor applications like toilets and laundry, ensuring potability for drinking and cooking, and building emergency preparedness. By exploring the versatility of collected rainwater, readers will appreciate this renewable resource and understand optimum techniques for different purposes.

Garden and Agricultural Irrigation

Gardening beds and agricultural fields are some of the best places to use harvested rainwater. Besides providing nutrients, rainwater has a few key advantages over tap or well water for irrigation needs. When used properly through well-planned systems and techniques, collected rainwater fully supports productive gardens and small farms.

Drip Irrigation Setup

One of the best ways to use collected rainwater in gardens is through a drip irrigation system. Unlike overhead sprinklers that waste water through evaporation, drip setup delivers water directly to root zones through a network of tubes, tapes, and emitters. Different drip kits suited for diverse garden beds, crops, and terrain are easily available. Careful planning is needed to lay out the tubing based on the garden layout to ensure full coverage without wastage. Installing appropriate filter assemblies at the irrigation supply line prevents emitter clogging from particles in the water.

Drip irrigation from rainwater tanks requires an automated watering system controlled by an electric timer. This regulates flow from the storage tank and ensures plants receive consistent moisture as per their watering needs. A pressure regulator also maintains optimal pressure for effective drip. Proper installation following the manufacturer's guidelines is important for the system to function efficiently with every drop of stored rainwater utilized by thirsty plants. With minimal maintenance, drip irrigation establishes a highly effective rainwater distribution process adapted for garden conditions.

Soaker Hoses for Beds and Raised Beds

Soaker hoses are a cost-effective alternative to rigid drip tapes that gently water plant bases through porous exteriors. They work very well for rows and blocks of vegetables or densely packed ornamental beds with various flowering annuals and perennials. Soaker hoses contour neatly along the edges or throughout beds, releasing water slowly where plant roots uptake moisture.

The hose is simply laid out beside or between plant rows, secured at intervals with U-shaped landscape staples. It connects to the rainwater supply line using adjustable clamp-on faucets that attach to rain barrels or above-ground tanks. Soaker hoses eliminate wasteful run-off by keeping water confined to root zones. Their flexibility makes them suitable for intricately curved beds without additional tubing joints or special fittings required for rigid drip lines. When garden designs change, soaker hoses can be easily repositioned or redistributed as compared to a fixed piped system.

Watering Cans for Individual Plants

While irrigation systems require initial layout and parts expenses, simple hand watering with durable cans practically serves scattered trees, shrubs, berry bushes, and potted vegetable plants or flowers stationed around gardens. A large, wide-mouthed watering can hold several gallons, reducing refill trips when filled from the rainwater collection system.

Using a can allows thorough moistening of the soil surface where plant roots uptake water and nutrients. This prevents fungal diseases caused by wet foliage while ensuring soils do not dry out between rains. Watering cans also enable individually regulating moisture for each plant's needs without disturbing others with excess drips or run-off, an advantage over blanket broadcast systems. The portable approach works well where permanent plumbing or electricity may not be readily feasible or economical for scattered plantings.

Selecting sturdy cans made of rust-resistant metals like aluminum makes them long-lasting garden tools. Ergonomic handles ensure comfortable carrying full distances between widespread plants. Marking max fill lines helps monitor the exact water volumes applied. Watering cans provide a simple solution for natural targeted irrigation from harvested rainwater, especially favored by gardening enthusiasts with small cultivation spaces.

Barrels on Wheels for Furry Crops

For expansive vegetable patches, cut flower crops, or small farm fields planted densely with furrowed vegetables, large rainwater-filled barrels mounted on sturdy wheels present a versatile irrigation method. Such field barrels hold 200-300 gallons; rolling them where needed across acres allows a bottom spigot to water furrows or soaker hoses laid out beneath rows.

Barrel wheels ease maneuvering heavy rainwater containers compared to hauling stationary tanks. They maintain consistent irrigation across wide beds of thirsty crops like melons, winter squash, or rows of leafy greens throughout the season.

Placing wheeled barrels at either end of furrows enables soaker hoses to be gravity-fed from the center. This spreads rainwater in both directions to thoroughly soak root zones. Marking barrel routes with landscape fabric minimizes compaction in narrow vehicle tracks. Choosing extra heavy-duty wheels rated to bear full barrel weights protects investments. Field barrels irrigate from harvested rainfall while preserving back strengths.

Soil Dampening for Greenhouses

Greenhouse crops needing steady moisture but lacking sophisticated drip systems can use collected rainwater for general soil dampening through alternate methods. Watering cans, trigger nozzle watering wands, or soaker hoses laid down greenhouse aisles help maintain an overall damp environment without overly wetting delicate potting mixes.

Recurring light douses raise the total moisture holding capacity for plant roots to uptake gradually between dampening. This prevents containers from drying out completely while also reducing the risks of drainage problems indoor garden settings pose. Greenhouse crops thriving with periodic light watering include salad greens, herbs, strawberries, and other moisture-loving plants.

Soaker hoses laid in alternate pathways allow covered areas to be dampened without damaging fragile crops by walking through. The permeable exterior of soaker hoses self-regulates flow to prevent flooding as water seeps uniformly into adjoining soils. Cleanup involves collapsing portable setups for storage till the next use.

Water Barrels for Orchards and Pastures

Large stationary tanks installed in orchards supply periodic soakings fruit and nut trees require to sustain floral buds, resulting in fruits that withstand dry seasons intact. Laying split flexible hoses from centrally placed tanks hydrates tree root zones thoroughly below canopies.

Pasture grasses similarly stay productive with deep intermittent watering from rainwater barrels or tanks. Placement in rolling grazing lands taps hills when needed to disperse rains falling where cattle water directly instead of trekking long distances. Alternately, portable troughs wheeled under shade trees quench livestock hungrier in summer. Constant access nourishes herds without depleting well reserves.

Periodic soil moisture probes help evaluate water retention to avoid over-wetting clay-rich patches more prone to compaction. Indicator plants like wilting blades signal when rains fill soil profiles for storage tanks to replenish through spillways. Orchards and pastures harness rain precisely as per fluctuating needs.

Indoor Uses: Toilets and Laundry

While rainwater meeting treatment can be used for drinking, many indoor non-potable uses are well-suited to collected but untreated water. A separate plumbing setup diverting rainwater to appropriate interior locations fulfills those needs sustainably. This section explores indoor rainwater applications in depth.

Plumbing for Toilet Flushing

One of the top indoor uses benefiting from rainwater diversions is toilet flushing. Municipal water allocated to this singular task amounts to a third of household consumption on average; by installing a dedicated graywater toilet system supplied via rainwater, significant potable water savings manifest.

Professional dual plumbing involves running separate above-ground piping to transport non-potable rainwater specifically for toilets. Strategically placed storage tanks buffer supply while electric-powered pumps drive flow on demand. Automated controls operate flush mechanisms at the push of a button, eliminating manual handling of non-potable water. Periodic system flushing and filter screening sustain optimal performance over the long run. Proper ventilation of enclosed tanks and pipes prevents internal buildup of microbial growth as well.

Laundry Needs

Apart from toilets, laundry presents another major indoor use excellently met through non-potable rainwater. Integrating designated rainwater plumbed to an external stub-out serves laundry tub needs without consuming a treated supply.

For full-capacity machines, the stub-out attaches to inlet valves using hose connectors. For hand washing, it supplies rainwater to a freestanding tub, sink, or repurposed barrel. This approach fulfills major water requirements of rinsing and sometimes washing clothes through prudent rainwater harvesting. Installation requires careful routing to place stub-outs in convenient laundry locations while avoiding potential cross-connections with potable lines. Pressure tanks assist in smooth operations.

Mopping and Floor Washing

Hard surfaces necessitate regular damp mopping or washing indoors as well. By setting up a mop sink filled solely through an outdoor rainwater spigot, such cleaning occurs without depleting treated water.

Locating the mop sink near entryways facilitates corralling dirt and debris for disposal with each task. Storing mop buckets and cleaning tools nearby maintains an organized workflow. Stainless steel or durable plastic construction withstands frequent use without rusting or warping over time. External spigots allow for refilling cleaned buckets and washing mop heads without tracking water indoors. Such practices preserve critical drinking water for human needs.

Water for Pets and Livestock

Quenching the thirsts of outdoor pets and livestock alike represents additional demands met through rainwater harvesting. Whether stabled animals or pets accommodated indoors, dedicated containers supplied by rainwater fulfill hydration needs conveniently without straining treatment systems.

For example, stationary automatic low-flow dog bowls keep family canine companions hydrated constantly via outdoor rainwater supply lines run to pet areas. Housed livestock equally benefit from through-wall feed troughs plumbed similarly to transport fresh rainwater. Portable containers also tap exterior spigots for pet water carry-ins as needed. Proper sanitization avoids contaminating indoor areas while pets gain access to pure water. Overall, harvested rainwater meets the critical moisture demands of all homestead animals humanely without depleting treated reserves.

Drinking and Cooking: Ensuring Portability

While rainwater fulfills numerous non-potable needs, utilizing it for internal consumption necessitates established treatment processes confirming absolute safety. Untreated collected rain carries microbial and chemical contaminants, rendering it unfit for drinking or food preparation without adequate disinfection and purification. This section details recommended methods.

Filtration and Disinfection

The most elementary approach to making rainwater potable involves multistage filtration followed by disinfection through approved techniques. Initial mechanical screening removes coarse debris and sediments, which can damage equipment. The finer particulate matter then gets filtered through layered particulate filters down to 5 microns.

Next, activated carbon filtration adsorbs potential pollutants like pesticides, herbicides, and inorganic chemicals, which may leach from rooftop materials or the atmosphere. Where chlorine residuals exist in rain, it removes these as well. Finally, disinfection with chlorine, ultraviolet light, or oxygen safely sterilizes water of microbiological impurities as per regulatory standards before internal use.

Portable sink-top or under-counter water treatment centers offer a plug-and-play option. Regular cartridge replacements maintain effectiveness. Larger centralized household or community treatment plants provide greater volumes for potable needs. Proper maintenance as per manufacturer guidelines ensures consistent, high-quality water production.

Water Testing and Monitoring

However, the highest standard for utilizing rainwater portably involves compliance with national drinking water guidelines enforced by periodic sampling and analysis. Guidelines specify maximum contaminant level thresholds for numerous water quality parameters that could impact public health.

Routine testing evaluates microscopic and chemical safety, including bacteria, nitrates, heavy metals, pH, and more. Sampling points consider potential post-collection recontamination risks. Automatic monitoring systems with sensors constantly assess treated water quality, sounding alarms for any treatment deficiencies detected. Documented test reports and an active notification program in case of failures demonstrate accountability.

For households, independent labs conduct analysis. Larger facilities require in-house testing equipment and certified testing personnel. Compiling periodic test data establishes a water quality history file for authorities' review. Correcting any inadequacies found and preventative equipment maintenance help sustain pristine potable water production reliant on rainwater as the primary source. Adhering to rigorous protocols safeguards community health.

Plumbing Identification

Proper identification distinguishes potable from non-potable plumbing networks maintained separately. Color-coding pipes, posting signs, and educational materials aid recognition. Potable water faucets remain clearly labeled to prevent cross-contamination. Double wall piping with air gaps between layers prevents back-siphonage of non-treated water into the potable system. Separately, room treatments encourage diligent maintenance of varied systems. Correct setups protect public welfare.

Rainwater Catchment Systems Design

Specialized design considerations apply to rainwater catchment systems supplying potable needs. As the primary source, collected water undergoes intensive pre-treatment and disinfection processes before being piped indoors.

Rooftop catchments utilize corrosion-resistant materials like galvanized steel, aluminum, concrete, or copper for durability and purity. Gutters include leaf screens and first-flush diverters, preventing contaminants in initial rains from entering storage. Overflows redirect excess non-potable water away from potable catchments.

Large-volume tanks maintain steady treated water flow and pressure. Where space is restricted, multiple smaller networked tanks evenly supply treatment systems. Buried structures require careful venting to prevent septic conditions. Tanks incorporate accessible manhole openings and cleaning ports for sanitization.

Pumps selected factor flow rates accommodating peak indoor and outdoor usage demands. Backflow preventers, bypass valves, dedicated treatment equipment, redundant disinfection, and monitoring safeguards uphold consistency. Regular replacement of consumables like filters ensures optimized functioning. Comprehensive designs factor in every contingency for impeccable potable water delivery.

Community Approvals

Utilizing rainwater as the primarily potable source requires conforming to all local codes and obtaining necessary approvals. Engagement helps address concerns through science-backed designs. Documentation includes schematics, treatment processes, testing protocols, intended uses, and emergency procedures.

Community rainwater utilities necessitate additional permits and regulatory oversight. Public health departments evaluate installation, operation, maintenance, and monitoring standards assured through comprehensive manuals. Periodic self-audits and open houses enhancing transparency aid social acceptance. Collaborations strengthen rainwater's role towards more resilient, independent, and climate-ready water supplies. Adhering to participatory processes empowers rainwater's potable potential.

Emergency Preparedness

Even with the utmost treatment reliability, contingencies prepare for potential supply interruptions. Back-up generators keep functioning during outages. Shelter-in-place portable purifiers process stored rainwater for survival. Dual verification alarms immediately alert operators to any anomalies.

Public awareness campaigns and printed advisories educate about boil water notices. Strategic reserve tanks hold interim supplies while rectifying rare failures. Cross-connections permit emergency backup from municipal mains if required. Comprehensive emergency response plans tested through drills ensure continuous access to potable water.

Emergency Preparedness

Beyond consistent usage, rainwater harvesting strengthens household and community resilience against disasters through strategic emergency stocks. This section details optimum preparations leveraging collected rainwater for survival needs during outages.

Stockpiling Non-Potable Rainwater

Storing ample non-potable rainwater caters to critical non-drinking needs if mains water loses pressure. Designated underground cisterns or elevated bolstered tanks maintain fixed volumes isolated from treatment systems.

Precautionary measures include screened overflow pipes and reinforced lids securing collected water from contamination. Tanks voluntarily restricted to graywater applications like flushing and cleaning avoid confusion amidst potability losses. Manual pumps or backup solar components keep non-potable supplies flowing. Stockpiles fulfill basic livelihood requirements while restoration works on treated networks or till alternative sources activate.

Potable Water Reservoirs

Back-up potable water reservoirs hold sufficient treated rainwater for limited drinking, food preparation, and medical needs if treatment disruption occurs. Buried concrete structures or enclosed ground-level tanks barrier collected water quality.

Redundant purification equipment like ultraviolet lamps replenished storage. Sensors constantly checking reserved water quality automatically trigger alarms at first signs of abnormality. Portable water testing kits enable rapid field checks. Such dedicated last-resort potable reservoirs tide over brief emergencies without comprising health.

Community Distribution

Large-scale centralized cisterns supply critical non-potable and potable water needs to stranded communities simultaneously. Strategically located long-duration storage tap surrounding roof catchments.

Pumping stations equipped to bypass non-functional treatment and draw directly from storage aid emergency response. Portable water buffaloes help transportation. Control centers coordinate rapid containment response with health and relief teams. Community rainwater stocks proving lifelines post disasters.

Shelter kits

Portable personal kits with manual pumps, water purification tablets, and collection and storage gear allow fallback rainwater self-sufficiency. Emergency education spreads such low-cost kit building.

Kits suit diverse needs – compact personal bottles for initial survival, collapsible containers gradually assembling rainwater stockpiles. Hand-powered ultraviolet wands sterilize limited stocks within bathtubs or sinks, extending potable reserves. Strategically placed group shelters utilize pre-installed catchments and first-flush diversion. Collected water meets cooking and sanitation needs within such refuges.

Crisis Communications

Crisis communications through multiple redundant channels update communities on emergency water availability, quality notices, and rationing as the situation evolves. Websites, sirens, and print advisories ensure timely guidance reaching the infirm and disadvantaged.

Emergency rainwater harvesting drills activating standalone catchments refine response logistics. Community emergency response teams trained in rainwater-based strategies strengthen preparedness. Collaborative planning optimizes harvested rains, serving critical needs through contingencies.

CHAPTER 10
MAINTENANCE, TROUBLESHOOTING, AND COMMON CHALLENGES

This chapter dives into the nitty-gritty of maintaining a rainwater harvesting system, troubleshooting common issues, and preparing for unexpected hiccups. Through the lens of a homesteader's tale, we explore the intricacies of ensuring the longevity and efficiency of your rainwater harvesting system.

Regular Maintenance Tasks

Regular maintenance tasks form the backbone of a thriving rainwater harvesting system, ensuring its longevity, efficiency, and the quality of the collected water. As a homesteader committed to sustainable living, engaging in routine upkeep becomes a ritual that transforms the rainwater harvesting system from a passive infrastructure to a dynamic, reliable water source.

Gutter Inspection and Cleaning

Gutter inspection and cleaning stand as the frontline defense against potential obstructions in the rainwater's journey from the collection surface to the storage tank. Imagine the gutters as the veins of the harvesting system, transporting the lifeblood of rainwater to where it's needed most. Over time, leaves, twigs, and other debris can accumulate, creating a barrier that impedes the flow. Homesteaders must cultivate the habit of regular gutter inspections, preferably at least twice a year, and clearing any debris. This simple yet crucial task not only prevents clogs but also safeguards against structural damage caused by overflowing water. It's a chore that connects the homesteader intimately with the ebb and flow of nature, reinforcing the symbiotic relationship between the harvesting system and the environment it relies upon.

Beyond the physical act of cleaning, gutter inspection involves a keen eye for signs of wear and tear. Rust, sagging, or misalignment could indicate issues that, if left unattended, might escalate into more significant problems. Regular inspections allow homesteaders to address these issues promptly, ensuring the long-term health of the entire rainwater harvesting infrastructure.

Filter Replacement and Cleaning

Filters play a pivotal role in maintaining the quality of harvested rainwater, acting as the gatekeepers that prevent impurities from entering the storage tank. Imagine them as the system's kidneys, purifying the collected rainwater before it becomes a vital resource for the homestead. However, these guardians are not immune to the wear and tear that comes with constant filtration. Regular inspection and cleaning of filters are paramount to ensuring their effectiveness and the overall health of the harvesting system. Homesteaders should establish a schedule for these tasks, aligning them with the unique characteristics of their environment and usage patterns.

During the inspection, homesteaders should carefully examine filters for any signs of damage, clogging, or degradation. A clogged filter not only compromises water quality but also strains the entire system, potentially leading to pump failures or decreased flow rates. Depending on the type of filter used, cleaning methods may vary, ranging from simple rinsing to more intricate processes.

Tank Inspection and Cleaning

The storage tank is the heart of the rainwater harvesting system, where the collected rainwater finds its sanctuary before being distributed for various uses. Regular tank inspections are akin to health check-ups, allowing homesteaders to identify and address potential issues before they escalate. This proactive approach not only safeguards water quality but also prevents costly repairs or replacements. The inspection process involves a meticulous examination of the tank's exterior and interior, focusing on structural integrity, signs of corrosion, and the presence of sediment or contaminants.

Homesteaders should pay particular attention to the structural components of the tank, checking for any signs of rust, dents, or leaks. Rust, often a result of prolonged exposure to the elements, can compromise the tank's structural integrity, leading to leaks or even failure. Promptly addressing rust issues with appropriate coatings or treatments is crucial for preventing further deterioration. Internally, tanks are susceptible to sediment buildup over time, especially in regions with high mineral content in the water. Regular cleaning and flushing of the tank's interior remove accumulated sediment, ensuring that the harvested rainwater remains pure and free from contaminants. In essence, tank inspection and cleaning are rituals of preservation, safeguarding not just the water within but the entire system's ability to function seamlessly.

Pump Maintenance

For those homesteaders utilizing a pump in their rainwater harvesting system, pump maintenance is a critical task to ensure the continuous flow of harvested rainwater. Picture the pump as the heart that propels the lifeblood of water through the veins of the system. Neglecting pump maintenance can lead to system failures, disrupting the very essence of self-sufficiency that homesteaders strive for. Regular checks and maintenance involve inspecting electrical connections, lubricating moving parts, and ensuring the pump's overall functionality.

Electrical components exposed to the elements are particularly vulnerable to wear and corrosion. Homesteaders should inspect wiring, connections, and switches, addressing any signs of damage promptly. Lubricating moving parts, such as bearings and motors, prevents friction and ensures the smooth operation of the pump. In regions experiencing extreme temperature fluctuations, the pump may be exposed to additional challenges. Installing protective measures, such as insulation or covers, can mitigate the impact of these environmental factors. By incorporating pump maintenance into the regular upkeep routine, homesteaders secure the reliable functioning of their rainwater harvesting systems, reinforcing their commitment to sustainable and self-reliant living.

Monitoring Water Levels

Regular monitoring of water levels is a practice that elevates rainwater harvesting from a passive endeavor to an active, responsive system. Imagine it as gauging the pulse of the homestead's water supply, allowing for timely adjustments in usage and conservation practices. This practice becomes especially crucial during periods of low rainfall or drought, where the availability of harvested rainwater may be limited. Homesteaders should establish a routine for checking water levels in their storage tanks, employing simple yet effective methods such as visual inspections or installing water level indicators.

Monitoring water levels enables homesteaders to make informed decisions about water usage, implementing conservation strategies when needed. This proactive approach prevents the risk of unexpectedly running out of water, a scenario that could compromise the self-sufficient lifestyle homesteaders aspire to.

Inspecting First Flush Diverters

First flush diverters are the gatekeepers at the entrance of the rainwater harvesting system, tasked with redirecting the initial rainfall—often laden with impurities—away from the main storage. Regular inspection and cleaning of these diverters are essential to maintain their effectiveness. Picture them as the system's first line of defense, ensuring that the water entering the storage tank is as pure as possible. Homesteaders should incorporate checks of first flush diverters into their overall maintenance routine, especially after significant rainfall events.

The inspection process involves ensuring that diverters are free from debris and that the mechanism functions smoothly. Debris, whether it be leaves, twigs, or other contaminants, can accumulate and compromise the diverter's ability to redirect the initial runoff effectively. Cleaning may involve simply removing debris or, in more sophisticated systems, disassembling the diverter for a thorough cleaning.

Addressing Common Issues: Algae, Mosquitoes, Leaks

In the intricate dance of rainwater harvesting, homesteaders encounter a myriad of challenges that test their commitment to sustainability. Among these challenges are the persistent foes of algae, mosquitoes, and leaks, which can disrupt the harmony of a well-established rainwater harvesting system. This section delves into the depths of each issue, providing insights and strategies to address these common adversaries effectively. By understanding the nature of these challenges and implementing proactive measures, homesteaders can maintain the purity of their collected rainwater and ensure the longevity of their harvesting systems.

Combatting Algae Growth

Algae, the green interloper, poses a constant threat to the integrity of harvested rainwater. As sunlight dances on the water's surface and nutrients abound, algae find a welcoming home. To combat this green invasion, homesteaders must consider a multi-faceted approach. First and foremost, the strategic placement of storage tanks plays a pivotal role; opting for opaque or tinted tanks prevents sunlight from penetrating and nourishing algae. Furthermore, the use of algaecides, carefully selected to be environmentally friendly, can be introduced to deter algae growth without compromising the ecological balance.

Shading emerges as a practical tool in the homesteader's arsenal against algae. Intelligently positioning structures or planting foliage strategically can cast protective shadows over tanks, impeding the relentless advance of algae. This natural shading not only curtails algae but also harmonizes with the homesteader's commitment to sustainability. Vigilant monitoring of tank conditions is equally crucial; periodic checks for any signs of algae growth allow for swift intervention before the issue escalates. Through a combination of thoughtful tank placement, eco-friendly interventions, and regular observation, homesteaders can thwart the encroachment of algae and preserve the quality of their harvested rainwater.

Mosquito Prevention Measures

The serene surface of rainwater in storage tanks can belie a potential threat – mosquitoes. Stagnant water becomes a breeding ground for these pesky insects, introducing the risk of water contamination and compromising the health of the entire harvesting system. Homesteaders must employ a multipronged approach to mitigate this concern. Introducing mosquito dunks or larvicides to the water ensures that mosquito larvae meet an early demise, preventing the emergence of the next generation. Additionally, sealing tanks effectively is paramount; ensuring all openings are fitted with fine-mesh screens acts as a physical barrier, preventing mosquitoes from infiltrating the system.

Proactive measures extend to the strategic placement of tanks. Elevating tanks off the ground and ensuring proper drainage around them minimize the chances of water pooling, a condition favorable to mosquito breeding. Homesteaders must also be vigilant in checking for any potential entry points, such as damaged screens or openings in the tank structure, which could serve as gateways for mosquitoes. A harmonious integration of preventive measures, from chemical interventions to thoughtful tank placement, safeguards the collected rainwater against mosquito infestations, aligning with the homesteader's commitment to sustainable, chemical-free water sources.

Detecting and Repairing Leaks

Leaks, the stealthy adversaries of rainwater harvesting systems, demand a vigilant eye and a proactive stance from homesteaders. Regular inspections form the frontline defense against potential leaks. Homesteaders should meticulously examine all components, including pipes, connectors, and tanks, for any signs of wear, corrosion, or damage. Early detection is crucial, as it allows for prompt intervention before a minor leak transforms into a major issue, potentially causing water wastage and structural damage.

The process of leak detection extends beyond the obvious visual checks. Homesteaders should actively listen for any unusual sounds, such as hissing or dripping, during system operation. Conducting these auditory inspections during and after rainfall events can reveal subtle leaks that may go unnoticed during dry periods. Additionally, pressure testing the system periodically can pinpoint leaks that might be challenging to identify through visual or auditory cues alone. Once a leak is detected, immediate repairs are imperative. Whether it involves patching a small hole or replacing a damaged component, addressing leaks promptly preserves water resources and sustains the overall functionality of the rainwater harvesting system. Through a combination of regular inspections, active listening, and systematic pressure testing, homesteaders can triumph over the stealthy challenge of leaks, ensuring the seamless flow and conservation of their harvested rainwater.

Winter Care and Freeze Protection

As winter blankets the homestead in a serene layer of snow, rainwater harvesting systems face a host of challenges. Winter care and freeze protection become paramount to sustain the functionality and integrity of these systems. In this detailed exploration, we navigate through the intricacies of winterizing your rainwater harvesting setup, ensuring that even in the coldest months, your self-sufficiency isn't compromised.

Insulating Above-Ground Components

In the frosty embrace of winter, above-ground components of rainwater harvesting systems are particularly susceptible to freezing. Pipes, exposed tanks, and other outdoor elements become vulnerable to the cold's grip, risking the formation of ice that can impede the system's functionality. Insulating above-ground components is a strategic and proactive measure to mitigate this risk. Utilizing materials designed for outdoor use, such as foam insulation or heat-reflective wraps, creates a thermal barrier, preventing the external temperature from impacting the water within the system. Proper insulation not only safeguards against freezing but also ensures the structural integrity of pipes and tanks, reducing the likelihood of cracks or other damage caused by temperature fluctuations.

Furthermore, careful attention to the installation of insulation is essential. Gaps or inadequately covered areas can compromise the effectiveness of the insulation, allowing cold air to infiltrate and create pockets of vulnerability. This aspect of winter care requires meticulous attention to detail, as a well-insulated system serves as a reliable defense against the challenges posed by freezing temperatures.

Heat Tracing for Pipes

In regions where winter's grasp is particularly unforgiving, the use of heat tracing for pipes emerges as a crucial strategy. Electric heat cables or tape, when strategically wrapped around pipes, provide a constant source of low-level heat, preventing water from freezing within the system. This method is especially effective in areas where temperatures frequently drop below freezing, safeguarding the flow of water, and preventing the formation of ice blockages in the pipes. Proper installation is essential to the efficacy of heat tracing, with attention to detail in securing the heating elements and ensuring coverage along the entire length of vulnerable pipes.

Integrating a thermostat into the heat tracing system adds an extra layer of intelligence. This allows the heat tracing to activate only when temperatures approach freezing, conserving energy and optimizing the efficiency of the winter care strategy. By adopting heat tracing for pipes, homesteaders can confidently face winter's challenges, knowing that their rainwater harvesting system is fortified against the threat of frozen pipes and potential disruptions to their water supply.

Utilizing Subsurface Tanks

The concept of subsurface tanks brings a natural advantage to winter care and freeze protection. By harnessing the insulating properties of the surrounding soil, subsurface tanks inherently resist the extremes of winter temperatures. This strategy involves placing the water storage tanks below ground level, where the earth acts as a thermal buffer, shielding the water from the full brunt of winter's cold. The soil's natural insulation properties create a more stable temperature environment, reducing the risk of freezing within the tanks.

The installation of subsurface tanks requires careful planning, considering factors such as soil composition, water table levels, and potential impacts on surrounding vegetation. Adequate drainage is also crucial to prevent water accumulation around the tanks, which could lead to unintended freezing. While subsurface tanks present unique challenges during installation, the long-term benefits in terms of winter care and freeze protection make them a viable and sustainable option for homesteaders facing harsh winter conditions.

Managing Snow Accumulation

As winter unfolds, the beauty of snowfall can present practical challenges for rainwater harvesting systems. Accumulated snow on tank lids, gutters, and other components can obstruct the collection of rainwater and potentially strain the structural integrity of these elements. Proactive management of snow accumulation is essential for maintaining the efficiency of the system. Regularly clearing snow from tank lids and gutters prevents the obstruction of collection surfaces, ensuring that the system continues to capture the available precipitation.

However, the removal of snow requires a delicate touch to avoid damaging the system. Abrasive tools or improper techniques can harm tank lids or damage gutters, compromising their functionality. Homesteaders must employ gentle methods, such as snow rakes or brooms, to clear snow without causing harm. This careful approach to managing snow accumulation is a crucial component of winter care, allowing the rainwater harvesting system to operate seamlessly even in the face of winter's frozen challenges.

Adjusting Water Usage During Freezing Conditions

An adaptive strategy in winter care involves adjusting water usage patterns during freezing conditions. Educating household members or users about the potential risks of freezing within the system and implementing water conservation practices can mitigate the impact of low temperatures. Simple measures, such as spreading water usage throughout the day to prevent long periods of inactivity in the system, can be effective in preventing freezing.

Additionally, communicating the importance of avoiding unnecessary water usage during extreme cold snaps ensures that the system isn't overtaxed. This proactive approach encourages a collective effort to adapt to the challenges of winter, fostering an understanding among users about the need for responsible water management during freezing conditions. By making these adjustments, homesteaders contribute to the overall resilience of the rainwater harvesting system in the face of winter's rigors.

Emergency Measures for Unexpected Freezing

Despite meticulous preparation, unexpected cold snaps can catch even the most diligent homesteader off guard. Having emergency measures in place is a crucial aspect of comprehensive winter care. Installing drain valves strategically allows for the controlled release of water, preventing it from freezing within the system. Temporary insulation solutions, such as blankets or thermal wraps, can be deployed quickly to shield vulnerable components during sudden drops in temperature.

Implementing emergency measures requires a proactive mindset and a thorough understanding of the rainwater harvesting system's vulnerabilities. Homesteaders should have a well-defined plan in place, including the location of drain valves and the availability of insulation materials. This preparedness ensures a swift response to unexpected freezing conditions, minimizing the potential impact on the system and preserving the integrity of the rainwater harvesting setup.

Expanding or Upgrading Your System

As the needs of a homestead evolve or technological advancements offer new possibilities, the prospect of expanding or upgrading a rainwater harvesting system becomes a consideration. This section explores the intricacies of expanding an existing system or upgrading its components to meet growing demands or leverage improved technologies.

Assessing Water Usage Needs

Before embarking on any expansion or upgrade, a thorough assessment of current and future water usage needs is essential. Consider factors such as the size of the homestead, the number of occupants, and potential changes in water demand. This evaluation forms the foundation for making informed decisions about the scale and scope of the expansion or upgrade.

Designing an Integrated Expansion

A well-designed expansion seamlessly integrates with the existing rainwater harvesting system. Consider factors such as the capacity of the current storage tanks, the efficiency of collection surfaces, and the distribution system. Ensuring synergy between new and existing components is key to maximizing the benefits of an expanded system.

Upgrading Filtration and Purification Systems

Technological advancements in filtration and purification offer opportunities to enhance the quality of harvested rainwater. Upgrading filtration systems or incorporating advanced purification methods can result in water that meets higher standards, expanding the range of potential uses. Evaluate the available technologies and choose upgrades that align with the desired water quality.

Exploring Smart Technologies

The era of smart technologies extends to rainwater harvesting. Consider incorporating sensors, monitoring systems, and automated controls to optimize the performance of the system. Smart technologies enable real-time monitoring, allowing homesteaders to track water levels, identify potential issues, and make data-driven decisions for efficient water management.

Scaling Up Storage Capacity

Increased water demand necessitates a corresponding increase in storage capacity. Assess the available space and explore options for adding larger tanks or additional storage units. Scaling up storage capacity ensures a reliable water supply, especially during periods of low rainfall.

Incorporating Redundancy for Reliability

Reliability is a cornerstone of any rainwater harvesting system. When expanding or upgrading, consider incorporating redundancy measures to ensure continuous functionality. This may involve redundant pumps, multiple storage tanks, or backup systems to mitigate the impact of component failures.

CHAPTER 11
SUSTAINABILITY AND ENVIRONMENTAL BENEFITS

In this chapter, we delve into the profound role of rainwater harvesting in conservation efforts, exploring its integration with other sustainable practices such as greywater usage and composting. Furthermore, we examine the broader implications of rainwater harvesting on local water systems and aquifers, envisioning a utopian world where this practice is embraced by all. By understanding the intricate connections between rainwater harvesting and sustainability, readers will be inspired to view their actions as integral components of a global movement toward ecological harmony.

The Role of Rainwater Harvesting in Conservation

Rainwater harvesting stands as a beacon of hope in the realm of sustainability, offering a tangible and impactful way for individuals and communities to contribute to environmental well-being. In this extensive exploration of the role of rainwater harvesting in conservation, we will delve into its multifaceted aspects, examining its contributions to landscape conservation, alleviation of stress on natural water sources, reduction of runoff and erosion, preservation of water quality, and its empowering impact on individual actions for global conservation.

Harnessing Rainwater for Landscape Conservation

One of the foundational elements of rainwater harvesting's contribution to conservation lies in its capacity to sustain landscapes without relying on external water supplies. The meticulous collection, storage, and distribution of rainwater transform it into a vital resource for nurturing gardens, crops, and green spaces. By doing so, rainwater harvesting not only mitigates the demand on municipal water systems but also actively participates in the preservation of local ecosystems. As individuals redirect rainwater to nourish their homesteads, they engage in a practice that embodies the delicate balance between human needs and environmental harmony. In this context, rainwater harvesting becomes a proactive step towards coexisting with nature, ensuring that the landscapes we inhabit thrive in harmony with the broader ecosystem.

Alleviating Stress on Natural Water Sources

Beyond sustaining individual landscapes, rainwater harvesting plays a pivotal role in alleviating stress on natural water sources. In many regions worldwide, rivers and aquifers face overexploitation, primarily due to escalating population demands. Rainwater harvesting, when adopted collectively, acts as a buffer against the depletion of these critical water sources. As communities reduce their reliance on overburdened rivers and aquifers, these natural water bodies gain the opportunity to regenerate and maintain a more balanced ecological equilibrium. This shift in water usage patterns is not merely a localized benefit; it becomes a contributory force toward the preservation of aquatic habitats and the biodiversity that depends on these ecosystems for survival.

Reducing Runoff and Erosion

Another facet of rainwater harvesting's conservation impact lies in its ability to mitigate the adverse effects of runoff and soil erosion. Traditionally, heavy rainfall leads to rapid runoff, carrying away valuable topsoil and introducing pollutants into nearby water bodies. Rainwater harvesting, with its emphasis on capturing and directing rainwater into storage systems, acts as a natural barrier against runoff. This not only safeguards fertile topsoil but also prevents the contamination of downstream water sources. The cumulative effect of countless individual rainwater harvesting systems paints a picture of landscapes more resilient to the erosive forces of nature. It becomes a harmonious dance between human intervention and natural processes, where rainwater, instead of becoming a destructive force, becomes a nurturing agent that sustains the vitality of the land.

Preserving Water Quality

Furthermore, rainwater harvesting contributes significantly to the preservation of water quality. Rainwater, in its purest form, is devoid of the contaminants commonly found in ground or surface water. By capturing rainwater before it comes into contact with rooftops and other surfaces, individuals can access a cleaner water source. This not only benefits the immediate users of harvested rainwater but also indirectly improves the overall water quality in the surrounding environment. It's a small yet impactful step towards creating a network of interconnected water systems that prioritize quality over quantity. As we embrace rainwater harvesting for its conservation potential, we simultaneously participate in a broader initiative to safeguard the purity of water, a fundamental element of life.

Empowering Individual Action for Global Conservation

Perhaps the most significant aspect of rainwater harvesting in the realm of conservation is its potential to empower individuals as active stewards of the environment. In a world where environmental challenges can seem overwhelming, the act of harvesting rainwater provides a tangible and achievable way for individuals to contribute to the larger cause of global conservation. By recognizing the interconnectedness of their actions with the broader environment, people can cultivate a sense of responsibility and agency, laying the foundation for a more sustainable future. Rainwater harvesting becomes a catalyst for a paradigm shift, where individuals no longer perceive themselves as passive consumers of resources but as active contributors to the resilience and vitality of the planet.

Fostering a Culture of Environmental Stewardship

The impact of rainwater harvesting on conservation extends beyond the physical realm into the cultural and societal domains. As individuals and communities embrace this practice, a culture of environmental stewardship begins to take root. The act of capturing and utilizing rainwater becomes a symbol of conscious living, signifying a commitment to sustainable practices. This cultural shift is not only a personal choice but a collective expression of values that prioritize ecological well-being. In this cultural evolution, rainwater harvesting becomes a unifying thread that weaves together diverse communities under the common banner of environmental responsibility.

Promoting Biodiversity through Sustainable Landscapes

A lesser-explored aspect of rainwater harvesting's role in conservation is its contribution to biodiversity through the creation of sustainable landscapes. As rainwater becomes a primary source for nurturing vegetation, these landscapes offer refuge and sustenance for a variety of flora and fauna. Native plants, adapted to the local climate, thrive with the chemical-free rainwater, attracting pollinators and creating habitats for diverse species. In this way, rainwater harvesting becomes a catalyst for promoting biodiversity at the micro-level, fostering ecosystems that support a rich tapestry of life. The conservation impact extends beyond water management; it becomes a holistic approach to creating environments that are in harmony with the intricate web of life.

Economic Resilience through Resource Independence

An often-overlooked dimension of rainwater harvesting's role in conservation is its contribution to economic resilience through resource independence. As individuals and communities rely on harvested rainwater for their water needs, they reduce their dependence on centralized water infrastructure. This resource independence translates into economic resilience, particularly in regions where access to traditional water sources may be unreliable or costly. By investing in rainwater harvesting systems, individuals not only contribute to conservation efforts but also establish a foundation for long-term economic sustainability, aligning environmental responsibility with financial prudence.

Enhancing Disaster Preparedness and Resilience

The conservation impact of rainwater harvesting extends to disaster preparedness and resilience. In times of water scarcity or natural disasters, individuals with rainwater harvesting systems are better equipped to weather the challenges. The stored rainwater becomes a crucial resource during periods of drought or when conventional water supplies are disrupted. This strategic reserve enhances the resilience of communities, ensuring that essential water needs are met even in adverse circumstances. In this way, rainwater harvesting becomes a cornerstone of disaster preparedness, offering a lifeline during times of environmental stress.

Educational Opportunities for Sustainable Living

The conservation journey facilitated by rainwater harvesting opens up rich educational opportunities for sustainable living. As individuals install, maintain, and optimize rainwater harvesting systems, they engage in a continuous learning process. This hands-on experience becomes a foundation for environmental education, fostering a deep understanding of water cycles, conservation principles, and sustainable practices. Beyond individual learning, rainwater harvesting becomes an educational tool in schools, community centers, and environmental organizations, inspiring the next generation to embrace a lifestyle that prioritizes conservation and responsible resource management.

Global Solidarity Through Conservation Awareness

The impact of rainwater harvesting on conservation transcends geographical boundaries, fostering a sense of global solidarity through conservation awareness. As communities around the world adopt and promote rainwater harvesting, a collective consciousness emerges. Individuals recognize that their efforts, no matter how small, contribute to a global movement for ecological harmony. This awareness becomes a unifying force, breaking down barriers and fostering a shared commitment to conservation. Rainwater harvesting becomes a symbol of shared responsibility, connecting diverse communities with the common goal of safeguarding the planet for present and future generations.

Combining with Other Sustainable Practices: Greywater, Composting, etc.

While rainwater harvesting alone is a potent tool for sustainable living, its effectiveness can be amplified when integrated with other eco-friendly practices. In this section, we explore the synergies between rainwater harvesting and complementary strategies like greywater usage, composting, permaculture, solar energy integration, natural building techniques, and more. By understanding how these practices harmonize, readers can envision a holistic approach to sustainability that extends beyond water conservation.

Synergy with Greywater Usage

Rainwater harvesting and greywater usage share a common goal – the efficient utilization of water resources. Greywater, which includes wastewater from domestic activities like laundry, bathing, and dishwashing, can be effectively integrated into rainwater harvesting systems. By combining the two, individuals create a comprehensive water management system that maximizes the use of both rainwater and greywater for non-potable purposes.

✓ Optimizing Water Use through Greywater Recycling

Greywater, when properly treated and recycled, can supplement rainwater for various household needs, such as watering plants, flushing toilets, and even certain cleaning tasks. This collaborative approach significantly reduces the demand for freshwater supplies for non-drinking purposes. Moreover, the integration of greywater recycling with rainwater harvesting fosters a closed-loop system, where water is used, treated, and reused within the confines of a homestead. This not only conserves water but also minimizes the environmental impact associated with wastewater disposal.

Synergy with Composting Practices

In the quest for holistic sustainability, the combination of rainwater harvesting and composting emerges as a dynamic duo. Composting, the natural decomposition of organic matter, yields nutrient-rich soil amendments that can significantly enhance the fertility of gardens and agricultural plots. When rainwater harvested from rooftops is channeled into composting areas, it accelerates the decomposition process and facilitates the production of nutrient-dense compost. This closed-loop approach transforms organic waste into a valuable resource, simultaneously reducing the reliance on synthetic fertilizers and promoting soil health.

✓ Reducing Environmental Footprint through Combined Practices

The integration of rainwater harvesting and composting creates a powerful synergy that extends beyond water conservation. Compost, rich in organic matter, acts as a natural soil conditioner, improving water retention and nutrient availability for plants. When rainwater is directed to composting areas, it aids in the breakdown of organic materials, fostering a more efficient composting process. This not only enhances the quality of harvested rainwater but also maximizes the benefits of compost for plant growth. The collaborative impact of these practices contributes to a substantial reduction in the overall environmental footprint, emphasizing the interconnectedness of sustainable actions.

Educational Synergies in Combined Practices

Furthermore, the combination of rainwater harvesting with other sustainable practices offers rich educational opportunities. Schools, community centers, and environmental organizations can use these integrated systems as educational tools to impart practical knowledge about water conservation, waste reduction, and sustainable living. This holistic approach not only enhances environmental literacy but also cultivates a sense of responsibility and stewardship among future generations, ensuring the perpetuation of sustainable practices.

✓ Community Building through Shared Sustainability

The collaborative nature of integrating rainwater harvesting with other sustainable practices extends beyond individual households. In communities where collective action is paramount, shared systems for rainwater harvesting, greywater recycling, and composting can be established. This communal approach not only maximizes efficiency but also fosters a sense of shared responsibility for environmental stewardship. As neighbors come together to implement and maintain these systems, a web of interconnected sustainability emerges, creating resilient communities that thrive on mutual support and shared values.

Synergy with Permaculture Principles

Permaculture, a design philosophy that emphasizes sustainable and regenerative practices, aligns seamlessly with rainwater harvesting. Integrating rainwater harvesting with permaculture principles enhances the overall resilience and productivity of a homestead or garden.

✓ Designing Water-efficient Permaculture Systems

Rainwater harvesting becomes an integral component of water-efficient permaculture designs. By capturing and storing rainwater, individuals can create microclimates within their permaculture systems, ensuring consistent water availability for diverse plantings. The stored rainwater serves as a buffer during dry spells, supporting the establishment of perennial crops and enhancing the overall biodiversity of the permaculture landscape.

✓ Promoting Closed-loop Systems

Incorporating rainwater harvesting into permaculture systems promotes closed-loop sustainability. The water collected from roofs and surfaces can be directed to swales, ponds, or keyline systems, facilitating efficient water distribution throughout the landscape. This not only conserves water resources but also minimizes the reliance on external inputs, creating a self-sustaining ecosystem where water, nutrients, and energy are intricately interconnected.

Integrating Solar Energy with Rainwater Harvesting

The integration of solar energy with rainwater harvesting presents a holistic approach to off-grid living and sustainable energy use. Harnessing the power of the sun alongside rainwater collection enhances the self-sufficiency and environmental impact of a homestead.

✓ Powering Water Pump Systems

Solar energy can be employed to power water pump systems used in rainwater harvesting. Photovoltaic panels can generate electricity to run pumps that transport harvested rainwater from storage tanks to distribution points. This synergistic approach ensures a continuous and sustainable supply of water without relying on conventional grid electricity.

✓ Sustainable Energy for Water Treatment

In off-grid settings, where water treatment is essential, solar energy can be harnessed to power purification systems. This ensures that rainwater collected is not only abundant but also meets quality standards. The combination of solar energy and rainwater harvesting exemplifies a harmonious integration of natural resources for essential household needs.

Harmonizing Natural Building Techniques with Rainwater Harvesting

Natural building techniques, which prioritize environmentally friendly and locally sourced materials, can be harmonized with rainwater harvesting to create homes that embody sustainability in both design and function.

✓ Green Roofs and Rainwater Harvesting Integration

Green roofs featuring vegetation, soil, and drainage layers seamlessly integrate with rainwater harvesting. The plants on green roofs utilize harvested rainwater, promoting biodiversity and providing additional insulation. This integrated approach not only reduces the environmental impact of construction but also enhances the energy efficiency of buildings.

✓ Sustainable Materials for Rainwater Collection Structures

When constructing rainwater collection structures, such as gutters and tanks, natural building materials like bamboo, recycled wood, or reclaimed metal can be employed. This ensures that the environmental impact of rainwater harvesting infrastructure remains minimal. The holistic synergy between natural building and rainwater harvesting contributes to the creation of homes that are not only aesthetically pleasing but also environmentally responsible.

Harmonizing Sustainable Agriculture Practices with Rainwater Harvesting

Sustainable agriculture practices, which prioritize environmental stewardship and long-term soil health, find a natural ally in rainwater harvesting. The combination of these practices promotes regenerative farming and resilient food production systems.

✓ Rainwater as a Resource for Sustainable Irrigation

In sustainable agriculture, water-efficient irrigation is paramount. Rainwater harvesting provides a reliable and chemical-free water source for irrigation needs. By integrating rainwater collection into agricultural systems, farmers can reduce their reliance on external water sources, contributing to the conservation of local water bodies.

✓ Enhancing Agroecological Diversity

The water collected through rainwater harvesting can be strategically used to support agroecological practices. Diverse crops and companion planting benefit from the consistent and localized water supply, promoting natural pest control and soil fertility. The synergy between rainwater harvesting and sustainable agriculture exemplifies a harmonious approach to food production that respects the ecosystem.

Integrating Smart Technologies for Enhanced Efficiency

The integration of smart technologies with rainwater harvesting systems brings a modern twist to sustainable living, enhancing efficiency and optimizing resource use.

✓ Sensors for Rainwater Collection Optimization

Smart sensors can be incorporated into rainwater harvesting setups to monitor rainfall patterns and optimize collection. These sensors provide real-time data, allowing individuals to adjust their harvesting systems based on weather forecasts. This ensures that rainwater is efficiently captured during periods of ample rainfall, maximizing storage capacity.

✓ Automated Distribution Systems for Water Efficiency

In conjunction with greywater usage, automated distribution systems can be employed to efficiently use rainwater for various household needs. Smart technologies enable precise control over water distribution, ensuring that harvested rainwater is utilized judiciously and reducing the reliance on municipal water for non-potable purposes.

Harmonizing Rainwater Harvesting with Forest Restoration

Forest restoration, a crucial element in combating deforestation and biodiversity loss, can be harmonized with rainwater harvesting to create landscapes that serve both ecological and human needs.

✓ Rainwater as a Catalyst for Forest Regeneration

In regions undergoing forest restoration efforts, harvested rainwater can be strategically directed to support tree planting and vegetation growth. This accelerates the natural regeneration of forests, providing a sustainable water source for newly planted trees during critical stages of establishment.

✓ Forest-Based Rainwater Harvesting Systems

In areas where forests are integral to water catchment, rainwater harvesting systems can be designed to complement and protect these ecosystems. Integrating rainwater collection with forest conservation ensures a harmonious balance between human water needs and the ecological services provided by healthy forests.

Enhancing Urban Agriculture with Rainwater Harvesting

As urban agriculture gains prominence, the integration of rainwater harvesting becomes instrumental in creating resilient and sustainable urban food systems.

✓ Rooftop Gardens and Urban Agriculture

Rooftop gardens, a hallmark of urban agriculture, can thrive with the integration of rainwater harvesting. The harvested rainwater provides a reliable and decentralized water source for cultivating fruits, vegetables, and herbs in urban settings. This localized approach to water supply contributes to the resilience of urban agriculture in the face of water scarcity.

✓ Sustainable Practices in Community Gardens

Community gardens, a vibrant aspect of urban agriculture, can adopt rainwater harvesting as a sustainable practice. By installing communal rainwater harvesting systems, urban communities can collectively contribute to water conservation, reduce dependence on external sources, and foster a sense of shared responsibility for sustainable living.

Impacts on Local Water Systems and Aquifers

Rainwater harvesting's effects on local water systems and aquifers are profound, creating a dynamic interplay between micro-scale practices and macro-scale environmental health. This section aims to delve deeper into the multifaceted impacts of rainwater harvesting on the intricate web of water sources that sustain communities and ecosystems. As we explore each facet, we uncover the transformative potential of rainwater harvesting in alleviating pressure on local water systems, recharging aquifers, improving water quality, mitigating urban heat island effects, enhancing biodiversity, and fostering community resilience in water-scarce regions.

Alleviating Pressure on Local Water Systems

Rainwater harvesting serves as a direct response to the increasing strain on local water systems, providing a sustainable alternative to traditional water sources. In water-scarce regions, where demand often outstrips supply, the widespread adoption of rainwater harvesting can alleviate pressure on municipal water supplies and natural water bodies. As individuals and communities reduce their reliance on centralized water sources, the cumulative effect becomes a distributed network of self-sufficiency. By capturing and utilizing rainwater for domestic and agricultural purposes, households actively participate in the conservation of local water resources. This decentralized approach not only lightens the burden on overtaxed water systems but also fosters a sense of responsibility among individuals to be active stewards of their local environment.

✓ Preserving Aquatic Ecosystems through Reduced Extraction

Beyond the immediate benefits of reduced demand on water systems, rainwater harvesting contributes to the preservation of aquatic ecosystems. In regions where rivers and streams face the threat of over-extraction, the adoption of rainwater harvesting practices represents a paradigm shift. By relying on harvested rainwater, individuals and communities decrease the need for large-scale extraction from natural water bodies. This reduction in extraction rates allows rivers and streams the opportunity to maintain healthier flow levels, which is essential for the well-being of aquatic habitats. As the pressure on these ecosystems lessens, the potential for biodiversity to thrive increases, creating a harmonious balance between human needs and the preservation of natural water sources.

Recharging Aquifers and Groundwater

The health of aquifers and groundwater reservoirs is critical for sustaining both rural and urban communities. Rainwater harvesting emerges as a natural mechanism for replenishing these underground water sources, offering a sustainable solution to the challenges of groundwater depletion. By capturing rainwater before it runs off into storm drains or evaporates, individuals contribute to the gradual recharge of aquifers. In regions heavily reliant on groundwater for drinking and agricultural needs, the widespread adoption of rainwater harvesting becomes a strategic tool for ensuring the long-term viability of these essential water sources.

✓ Sustainable Recharge Practices for Aquifers

The sustainable recharge of aquifers through rainwater harvesting involves not only the quantity of water but also the quality. As rainwater percolates into the ground, it undergoes a natural filtration process, removing impurities and contaminants. This filtered water replenishes the aquifers with cleaner, more pristine water compared to water from other sources. The cumulative effect of widespread rainwater harvesting, with its dual emphasis on quantity and quality, becomes a cornerstone for maintaining healthy groundwater reservoirs. In this way, rainwater harvesting not only addresses immediate water needs but also contributes to the overall resilience and sustainability of underground water sources.

Improving Water Quality in Local Ecosystems

The quality of water in local ecosystems is intricately linked to the practices of the surrounding human population. Rainwater harvesting plays a pivotal role in improving water quality by minimizing the introduction of pollutants and contaminants into natural water bodies. As rainwater is collected from clean surfaces, it inherently carries fewer impurities compared to water sourced from urban runoff or industrial discharges. This reduction in pollution levels has cascading effects on the health of rivers, streams, and other water habitats, creating a positive feedback loop of environmental well-being.

✓ Reducing Contaminant Influx through Urban Runoff

Urban areas, with their expansive impervious surfaces, often contribute to the contamination of local water bodies through runoff. Rainwater harvesting interrupts this cycle by intercepting rainwater before it reaches the ground and directing it into storage systems. As a result, the pollutants that typically accompany urban runoff, such as heavy metals, oils, and chemicals, are prevented from entering local ecosystems. The act of capturing rainwater becomes a form of natural stormwater management, minimizing the detrimental impact of urbanization on water quality. This reduction in contaminant influx not only benefits aquatic life but also safeguards the health of downstream communities reliant on these water sources.

Mitigating Urban Heat Island Effects

Urban heat islands, characterized by elevated temperatures in urbanized areas, pose significant challenges to both the environment and public health. Rainwater harvesting emerges as a strategic tool for mitigating the urban heat island effect, offering a localized solution that contributes to the overall cooling of urban spaces. As rainwater is captured on rooftops and other surfaces, it serves as a natural coolant, dissipating heat and reducing the temperature of buildings and their surroundings. This localized cooling effect, when multiplied across a community, becomes a powerful means of creating more habitable and sustainable urban environments.

✓ Cooling Urban Environments through Distributed Rainwater Capture

The cooling effect of rainwater harvesting is particularly pronounced in densely populated urban areas with high concentrations of impervious surfaces. By capturing rainwater at the source, individuals and communities actively participate in a distributed network of heat reduction. The stored rainwater on rooftops not only prevents immediate surfaces from absorbing excessive heat but also contributes to the moderation of air temperatures in the surrounding area. This collaborative effort in cooling urban environments creates pockets of respite from the heat, fostering more livable and environmentally conscious urban landscapes.

Enhancing Biodiversity in Water-dependent Habitats

Biodiversity in water-dependent habitats is intricately connected to the health and quality of water sources. Rainwater harvesting, by positively impacting local water systems, becomes a catalyst for enhancing biodiversity in rivers, streams, wetlands, and other water-dependent ecosystems. As the pressure on natural water sources lessens, these habitats gain the resilience needed to support a diverse array of flora and fauna, contributing to the overall ecological balance.

✓ Restoring Ecological Balance through Reduced Stress on Water Sources

The stress on water-dependent habitats is often a consequence of excessive extraction and pollution. Rainwater harvesting addresses these challenges by reducing stress on natural water sources and creating conditions conducive to the flourishing of aquatic ecosystems. Healthy rivers and streams, supported by sustained rainwater harvesting practices, become havens for a variety of species. The restoration of ecological balance in these habitats is not only a testament to the positive impacts of rainwater harvesting but also underscores its role in fostering a harmonious coexistence between human activities and the natural world.

Community Resilience in Water-scarce Regions

Water scarcity presents a formidable challenge in many regions, necessitating innovative and sustainable solutions. Rainwater harvesting, when embraced at the community level, becomes a cornerstone for building resilience in water-scarce regions. As individuals and communities collectively adopt rainwater harvesting practices, they create a buffer against the uncertainties of water shortages, ensuring a reliable water supply for essential needs. This communal resilience goes beyond immediate water availability, fostering a sense of unity and shared purpose in the face of environmental challenges.

✓ Collective Water Security through Community-wide Adoption

In water-scarce regions, the collective adoption of rainwater harvesting transforms communities into architects of their own water security. By harvesting and storing rainwater on a community scale, individuals safeguard against the fluctuations in traditional water sources. This communal approach not only ensures a consistent water supply but also reduces the reliance on centralized water infrastructure, which may be vulnerable to disruptions. The act of coming together for a shared cause – securing a sustainable water future – fosters a sense of collective responsibility and solidarity, laying the groundwork for resilient communities in the face of changing environmental dynamics.

The Role of Technology in Optimizing Rainwater Harvesting

Technological advancements play a crucial role in optimizing rainwater harvesting systems, making them more efficient, user-friendly, and adaptable to diverse environments. From innovations in harvesting mechanisms to sophisticated filtration systems, technology empowers individuals and communities to harness rainwater with greater precision and effectiveness. This section explores the evolving landscape of rainwater harvesting technology, showcasing how these advancements contribute to the broader goals of water conservation and sustainability.

✓ Innovative Harvesting Mechanisms for Maximum Efficiency

Modern rainwater harvesting systems often incorporate innovative mechanisms to maximize efficiency in capturing and storing rainwater. These may include advanced rooftop collection systems, smart sensors to monitor rainfall patterns, and automated storage solutions. Innovations in harvesting mechanisms not only enhance the quantity of harvested rainwater but also streamline the entire process, making it more accessible to a broader audience. As technology continues to evolve, the potential for even more sophisticated and user-friendly harvesting systems holds promise for the widespread adoption of rainwater harvesting practices.

✓ Smart Filtration and Purification Systems for Quality Assurance

Ensuring the quality of harvested rainwater is paramount for its diverse applications, from irrigation to domestic use. Technological advancements in filtration and purification systems have significantly improved the ability to maintain high water quality standards. Smart filtration systems equipped with sensors and automated purification processes ensure that the collected rainwater meets established safety guidelines. This not only broadens the scope of applications for harvested rainwater but also instills confidence in users regarding its safety and reliability. The intersection of technology with rainwater harvesting is not just about quantity; it's about optimizing every aspect of the process to achieve a harmonious balance between efficiency and environmental responsibility.

Regulatory Considerations and Policy Implications

The adoption of rainwater harvesting practices necessitates a supportive regulatory framework and clear policy guidelines to ensure its effective implementation. This section explores the regulatory considerations and policy implications associated with rainwater harvesting, examining how governments and local authorities can play a pivotal role in promoting and sustaining this environmentally beneficial practice.

✓ Legal Frameworks Encouraging Rainwater Harvesting

Creating an environment conducive to rainwater harvesting requires the establishment of legal frameworks that incentivize and support individuals and communities. Governments can introduce policies that provide financial incentives, tax breaks, or rebates for installing rainwater harvesting systems. Clear legal frameworks also address potential concerns regarding water rights, property rights, and safety standards associated with rainwater harvesting. By enacting legislation that encourages and supports rainwater harvesting, policymakers contribute to the broader goals of water conservation and sustainable resource management.

✓ Local Ordinances and Building Codes for Integration

At the local level, ordinances and building codes play a crucial role in integrating rainwater harvesting into the fabric of communities. Local authorities can mandate the inclusion of rainwater harvesting systems in new construction projects or offer streamlined approval processes for retrofitting existing structures. The incorporation of rainwater harvesting requirements into building codes ensures that this sustainable practice becomes an integral part of urban and rural development. Through thoughtful local ordinances, policymakers create an environment where rainwater harvesting is not just a choice but a standard practice for responsible resource management.

Economic Considerations and Cost-Benefit Analysis

The economic aspects of rainwater harvesting are pivotal considerations for individuals, communities, and policymakers. This section delves into the economic considerations associated with rainwater harvesting, exploring the initial costs, long-term benefits, and the broader economic impact of widespread adoption.

✓ Initial Costs and Return on Investment

The upfront costs of installing rainwater harvesting systems can be a significant consideration for individuals and communities. This includes expenses for harvesting infrastructure, storage tanks, filtration systems, and installation. However, it's crucial to view these costs through the lens of long-term benefits and return on investment. Rainwater harvesting systems, once installed, contribute to ongoing water savings, reduced utility bills, and potential financial incentives. A comprehensive cost-benefit analysis reveals that the initial investment pays off over time, making rainwater harvesting an economically viable and sustainable choice for resource-conscious individuals and communities.

✓ Job Creation and Economic Stimulus

The widespread adoption of rainwater harvesting has implications beyond individual households, extending to the broader economy. The installation, maintenance, and innovation associated with rainwater harvesting systems create employment opportunities in various sectors. From skilled technicians and engineers involved in system installation to researchers contributing to technological advancements, rainwater harvesting becomes a source of job creation and economic stimulus. Policymakers can leverage this aspect by promoting training programs, research initiatives, and economic incentives that further bolster the economic benefits associated with rainwater harvesting.

Education and Outreach for Sustainable Practices

Educating communities about the benefits and practices of rainwater harvesting is a crucial component of its widespread adoption. This section explores the role of education and outreach in promoting sustainable rainwater harvesting practices, emphasizing the need for awareness campaigns, educational programs, and community engagement initiatives.

✓ Public Awareness Campaigns for Behavior Change

Public awareness campaigns play a pivotal role in fostering behavior change and encouraging the adoption of rainwater harvesting practices. These campaigns can utilize various media channels, including social media, television, and community events, to disseminate information about the environmental benefits, economic advantages, and ease of implementation associated with rainwater harvesting. By creating a narrative that connects rainwater harvesting to broader themes of sustainability and environmental responsibility, public awareness campaigns become catalysts for cultural shifts towards more water-conscious lifestyles.

✓ Incorporating Rainwater Harvesting into Educational Curricula

Integrating rainwater harvesting into educational curricula at various levels contributes to building a generation that values sustainability. Schools and universities can incorporate practical lessons on rainwater harvesting, involving students in the design, installation, and maintenance of harvesting systems. This hands-on approach not only imparts technical knowledge but also instills a sense of responsibility and environmental stewardship in future generations. Education becomes a transformative tool, shaping the attitudes and behaviors of individuals toward a more sustainable and water-responsible future.

CHAPTER 12
COMMUNITY AND EDUCATIONAL OUTREACH

In a small town nestled between rolling hills and fertile valleys, a community embarked on a journey that would not only transform their immediate environment but also sow the seeds of sustainable change for generations to come. This is a tale of neighbors turning into collaborators, of shared visions and collective efforts, all centered around the simple yet profound act of harvesting rainwater. It began with a few passionate individuals who saw the potential for positive change in a resource as commonplace as rain. As the droplets fell from the heavens, so did the realization that they could be harnessed to nourish not only individual homesteads but the entire community.

Building a Community Harvesting Project

Understanding Community Dynamics

Communities are intricate ecosystems of human interactions, shared values, and interdependence. Before embarking on the technicalities of a rainwater harvesting project, it is imperative to delve into the nuanced dynamics that define a community. Identifying key stakeholders and understanding existing social structures lay the groundwork for successful collaboration. Are there community leaders whose endorsement could significantly sway the project's reception? Are there pre-existing communal activities that could be integrated into the rainwater harvesting initiative? By conducting thorough community analyses, potential pitfalls and opportunities can be anticipated, fostering a more cohesive and effective project.

Navigating community dynamics also involves recognizing and respecting cultural nuances. Every community has its unique identity, shaped by traditions, values, and histories. Integrating rainwater harvesting into the fabric of daily life requires an appreciation of these cultural aspects. It might involve drawing parallels between traditional water conservation practices and modern rainwater harvesting methods, ensuring that the initiative aligns seamlessly with the community's identity. By acknowledging and incorporating these elements, the project becomes not just an infrastructural intervention but a reflection of the community's values and heritage.

Mapping Water Usage Patterns

The success of any rainwater harvesting project hinges on a comprehensive understanding of the community's water consumption habits. This involves conducting surveys, engaging in open dialogues, and collaborating with local authorities to gather data on peak water consumption times and areas with the highest demand. Mapping water usage patterns is not only an exercise in data collection but a strategic tool for designing efficient rainwater harvesting systems.

Identifying peak consumption times allows for the optimization of rainwater collection and storage capacities to meet the community's needs during high-demand periods. Additionally, understanding the geographical distribution of water usage informs the strategic placement of harvesting systems. For example, areas with extensive gardening might benefit from decentralized rain barrels, while communal spaces like parks might require more substantial catchment solutions. This level of detail ensures that the rainwater harvesting project is not only effective but tailored to the specific needs and behaviors of the community, maximizing its impact.

Selecting Appropriate Harvesting Systems

Communities come in diverse shapes and sizes, and so should their rainwater harvesting systems. This subsection explores the myriad options available, ranging from simple rain barrels to more sophisticated catchment and storage solutions. Each system has its advantages and limitations, and the selection process involves a careful consideration of various factors.

Factors such as climate play a pivotal role. In arid regions, where rain is infrequent, systems with larger storage capacities may be necessary to sustain the community through prolonged dry spells. Conversely, in regions with frequent rainfall, smaller and more distributed systems might be sufficient. The terrain is another crucial consideration; communities in hilly areas may benefit from gravity-based systems, while those in flat terrains might require pumps for effective water distribution. The size and density of the community also influence the choice of systems, with larger communities often requiring more complex infrastructure. Striking the right balance between cost-effectiveness and sustainability is key to selecting systems that align with the unique needs of the community.

Legal and Regulatory Considerations

Navigating the legal landscape is a vital aspect of implementing a community-based rainwater harvesting project. This involves understanding and adhering to local regulations governing water usage and conservation. The legal considerations extend beyond obtaining necessary permits; they also encompass addressing potential concerns and mitigating risks associated with the project.

Community leaders and project organizers need to collaborate with local authorities to ensure that the rainwater harvesting initiative aligns with existing regulations. This might involve seeking exemptions or advocating for supportive policies that encourage sustainable water practices. Additionally, addressing potential concerns from community members or stakeholders is crucial. This could include addressing fears of water scarcity, property damage, or disruptions to existing water supply systems.

By proactively engaging with legal and regulatory considerations, the community project not only becomes environmentally sound but also gains the necessary support and endorsement from local authorities.

Funding and Resource Mobilization

The financial aspect of a community rainwater harvesting project is often a determining factor in its feasibility and success. This subsection explores the diverse avenues for funding, from government grants and subsidies to community crowdfunding and partnerships with local businesses. Beyond monetary considerations, it also delves into mobilizing local resources, fostering a sense of communal ownership and pride in the project.

Government grants and subsidies can significantly alleviate the financial burden of the project, and understanding the application processes and eligibility criteria is essential. Community crowdfunding, on the other hand, not only raises funds but also builds a sense of collective investment and responsibility. Exploring partnerships with local businesses can provide not only financial support but also access to materials and expertise. Mobilizing local resources involves tapping into the skills and talents within the community, transforming the project into a collaborative effort. By diversifying funding sources and incorporating community resources, the rainwater harvesting initiative becomes a shared endeavor, reinforcing the notion of collective responsibility.

Sustaining Momentum: Maintenance and Community Engagement

Implementing a community rainwater harvesting project is not a one-time endeavor; it requires ongoing commitment. This section delves into the critical components of maintenance, from regular inspections to addressing potential issues promptly. Furthermore, fostering continuous community engagement ensures that the project remains relevant and ingrained in the collective consciousness.

Ensuring Effective Maintenance Practices

The longevity and effectiveness of rainwater harvesting systems depend on meticulous maintenance. Regular inspections of gutters, filters, and storage facilities are imperative to identify and rectify issues before they escalate. This involves creating a maintenance schedule, appointing responsible individuals within the community, and providing training on basic upkeep. Additionally, incorporating smart technologies, such as sensors that monitor water quality and system functionality, can streamline the maintenance process, ensuring the longevity of the infrastructure.

Addressing Issues Promptly

Issues are inevitable in any infrastructure project, and rainwater harvesting is no exception. This subsection emphasizes the importance of addressing issues promptly to prevent cascading problems. Whether it's a clogged filter, a leaky storage tank, or damage to distribution pipes, swift action is essential. Establishing a community response team or collaborating with local technicians ensures that problems are identified and resolved in a timely manner. This not only preserves the integrity of the rainwater harvesting systems but also fosters a sense of reliability and trust within the community.

Community Engagement as a Catalyst for Sustainability

The success of a community rainwater harvesting project extends beyond the physical infrastructure; it is intricately tied to community engagement. This involves fostering a sense of ownership and pride in the project, making it an integral part of the community's identity. Regular community meetings, newsletters, and social events centered around the project create opportunities for residents to share their experiences, voice concerns, and celebrate achievements. The project becomes a symbol of collective action, reinforcing the notion that sustainability is a shared responsibility.

Educating Neighbors and Local Schools

Education is the linchpin of sustainable change, and extending the knowledge and benefits of rainwater harvesting to neighbors and local schools transforms individuals into catalysts for a broader cultural shift toward environmental consciousness. This section provides guidance on effectively disseminating information, engaging with different demographics, and inspiring the next generation of eco-conscious leaders.

Crafting Educational Materials

Effective communication is pivotal in fostering understanding and enthusiasm within the community. The key is to present information in accessible and engaging formats that resonate with diverse learning styles. Clear and visually appealing materials can simplify complex concepts related to rainwater harvesting, making the information more digestible for a broad audience.

Creating educational materials also involves tailoring content to address specific community concerns and interests. For example, if water scarcity is a prevalent issue, the materials can highlight how rainwater harvesting directly addresses this challenge. If the community has a strong gardening culture, the emphasis might be on the benefits of chemical-free rainwater for plants. By aligning educational materials with the unique needs of the audience, individuals are more likely to connect with the information and, subsequently, adopt rainwater harvesting practices.

Neighborhood Workshops and Information Sessions

Transforming knowledge into action requires personal engagement. By demystifying the process of rainwater harvesting, individuals are more likely to embrace and implement these practices in their own homes.

Workshops can take various forms, from practical demonstrations of setting up rain barrels to discussions on the environmental impact of rainwater harvesting. The key is to make the sessions interactive, allowing participants to actively engage with the material. Incorporating success stories from within the community can also inspire neighbors, showcasing the tangible benefits of rainwater harvesting. Additionally, providing resources and contacts for further assistance ensures that individuals feel supported in their journey toward sustainable water practices.

Collaborating with Local Schools

Schools are not just centers of education; they are influential hubs within the community. Integrating rainwater harvesting into the school curriculum offers a structured and systemic approach to education.

Collaboration can take various forms, from guest lectures by community leaders involved in rainwater harvesting projects to incorporating hands-on activities into science and environmental studies classes. Organizing field trips to community rainwater harvesting sites provides students with real-world exposure, reinforcing theoretical concepts. Moreover, involving students in practical projects, such as setting up rain barrels on school premises, instills a sense of responsibility and ownership. The ripple effect extends beyond the classroom, influencing families and creating a generation that views water conservation as a natural and essential practice.

Engaging with Diverse Demographics

Communities are diverse, comprising individuals with varying levels of awareness and interest in environmental issues. Strategies for reaching out to different age groups, socio-economic backgrounds, and cultural contexts are explored, fostering inclusivity in the journey toward sustainable living.

Engaging with diverse demographics involves understanding the unique needs and challenges faced by different groups within the community. For example, elderly residents might be more concerned with the practicality and ease of maintenance of rainwater harvesting systems. Low-income households might require information on cost-effective solutions and potential financial assistance. Tailoring communication to resonate with specific cultural values and traditions ensures that the message is received and embraced. By acknowledging and addressing the diversity within the community, educational outreach becomes a unifying force that transcends boundaries.

Measuring Impact and Adapting Strategies

The effectiveness of educational outreach lies in its impact. From tracking the adoption of rainwater harvesting practices to assessing changes in community attitudes toward water conservation, understanding the impact enables continuous improvement.

Measuring impact involves both quantitative and qualitative assessments. Quantitatively, tracking the number of households adopting rainwater harvesting, the volume of water saved, and the reduction in reliance on external water sources provides tangible metrics. Qualitatively, gathering feedback through surveys, interviews, and community meetings gauges the perceived benefits and challenges. This dual approach allows for a comprehensive understanding of the initiative's influence on both individual behaviors and the community's overall mindset.

Adapting strategies based on feedback is an integral part of the educational outreach process. If certain demographics express resistance or encounter barriers, adjustments can be made to address specific concerns. Likewise, if certain communication channels prove more effective than others, reallocating resources accordingly maximizes impact.

Creating a Network of Sustainability Advocates

Beyond individual households, this segment explores the potential of creating a network of sustainability advocates within the community. Empowering individuals to become ambassadors for rainwater harvesting expands the reach of the initiative. This network can collaborate on broader environmental issues, creating a web of influence that extends beyond water conservation.

Building a network of sustainability advocates involves identifying and nurturing individuals passionate about environmental issues. This might involve organizing community forums where like-minded individuals can connect and share ideas. Providing resources and training for advocates equips them to effectively communicate the benefits of rainwater harvesting and broader sustainability practices. The network can then collaborate on community-wide initiatives, such as tree planting campaigns or waste reduction projects, creating a holistic approach to environmental stewardship. By fostering a sense of shared responsibility, communities can collectively navigate the challenges of sustainability in a rapidly changing world.

CHAPTER 13
BEYOND THE BASICS

This chapter goes beyond conventional techniques, introducing advanced filtration systems, smart monitoring with digital tools, incorporating solar pumps, and optimizing landscape design for superior rainwater utilization.

Advanced Filtration Systems

In the realm of rainwater harvesting, filtration stands as a critical process to ensure the collected water meets the highest standards of purity. Basic filtration systems are essential, but for those seeking a more refined and advanced approach, a spectrum of innovative filtration technologies exists.

High-Efficiency Mesh Filters

The foundation of advanced filtration lies in high-efficiency mesh filters. These filters surpass conventional screens, utilizing advanced materials that sieve out even the finest particulate matter. Their design ensures minimal clogging, leading to consistently clear and contaminant-free rainwater. Homesteaders venturing into precision agriculture or those with a focus on water quality for domestic use will find high-efficiency mesh filters to be an invaluable addition to their rainwater harvesting system.

Activated Carbon Filtration

Taking filtration to the next level, activated carbon filtration proves to be a game-changer in purifying rainwater. Activated carbon has a remarkable ability to adsorb impurities and contaminants, ranging from organic compounds to chlorine and volatile organic compounds (VOCs). This advanced filtration method not only enhances the taste and odor of collected rainwater but also ensures its suitability for a broader spectrum of applications, including drinking and cooking.

Reverse Osmosis Systems

For those aiming for the pinnacle of water purification, integrating reverse osmosis (RO) systems into rainwater harvesting setups provides unparalleled results. RO systems employ a semi-permeable membrane to remove microscopic impurities, ensuring a level of purity that rivals distilled water. While energy-intensive, RO systems offer an unmatched solution for homesteaders with stringent water quality requirements or those exploring rainwater as a potable water source.

Nanotechnology in Filtration

Pushing the boundaries of innovation, nanotechnology has found its application in rainwater filtration. Nanoparticles possess unique properties that enhance filtration efficiency, capturing particles at a molecular level. This level of precision ensures the removal of contaminants that might escape traditional filtration methods.

Ultraviolet (UV) Sterilization

Complementing filtration with disinfection, UV sterilization has gained prominence in advanced rainwater harvesting systems. UV light effectively neutralizes bacteria, viruses, and other pathogens present in collected rainwater, providing an additional layer of protection. Homesteaders concerned about waterborne diseases or aiming for medical-grade water quality will find UV sterilization to be an indispensable component in their advanced rainwater harvesting toolkit.

Smart Filtration Integration

In the era of smart technologies, integrating sensors and monitoring systems into filtration setups enhances efficiency and reduces maintenance efforts. Smart filtration systems can automatically detect filter clogs, monitor water quality in real-time, and even initiate self-cleaning processes when necessary. This intersection of technology and water management elevates rainwater harvesting to a new echelon, catering to the demands of modern, tech-savvy enthusiasts.

Smart Monitoring with Digital Tools

The advent of digital technology has revolutionized the way we interact with and manage our surroundings, and rainwater harvesting is no exception. Smart monitoring systems, equipped with digital tools and sensors, offer a quantum leap in efficiency, control, and data-driven decision-making for rainwater enthusiasts.

Wireless Sensor Networks

The cornerstone of smart monitoring in rainwater harvesting lies in the deployment of wireless sensor networks. These networks consist of strategically placed sensors that continuously collect data on various parameters such as rainfall intensity, water level in storage tanks, and overall system performance. By wirelessly transmitting this information to a central hub, enthusiasts gain real-time insights into their rainwater harvesting setup, enabling proactive adjustments and optimizations.

Cloud-Based Data Analytics

Harnessing the power of cloud computing, rainwater harvesting enthusiasts can now leverage sophisticated data analytics tools to derive meaningful insights from the vast amount of data generated by their systems. Cloud-based platforms enable the storage, processing, and analysis of data in real-time, allowing enthusiasts to make informed decisions regarding water usage, system efficiency, and predictive maintenance.

Smartphone Applications for Remote Monitoring

In an era where smartphones have become an extension of our daily lives, dedicated applications for rainwater harvesting add an unprecedented level of convenience and accessibility. These apps connect to the smart monitoring system, providing enthusiasts with real-time updates, alerts, and even remote control over certain aspects of their rainwater harvesting setup. Whether checking water levels while away from home or adjusting filtration parameters with a few taps on a smartphone screen, these applications empower enthusiasts to be in complete control of their water management systems.

Machine Learning Algorithms for Predictive Analysis

As data accumulates from smart monitoring systems, machine learning algorithms come into play, offering predictive analysis capabilities. These algorithms can forecast rainfall patterns, predict system malfunctions before they occur, and optimize filtration and storage processes based on historical data. The integration of machine learning into rainwater harvesting not only enhances efficiency but also contributes to a more sustainable and resource-conscious approach.

Internet of Things (IoT) Integration

The Internet of Things (IoT) has transcended conceptual boundaries to become a reality in rainwater harvesting. IoT devices, ranging from smart valves to connected rain gauges, enable seamless communication and coordination within the rainwater harvesting system. Enthusiasts can create a fully interconnected ecosystem where each component communicates with others, ensuring optimal performance and responsiveness to changing environmental conditions.

Augmented Reality (AR) for System Visualization

Innovations in augmented reality bring a new dimension to rainwater harvesting, allowing enthusiasts to visualize their entire system in real-time. AR applications overlay digital information onto the physical world, enabling enthusiasts to see water flow pathways, tank levels, and system components through the lens of a smartphone or AR headset. This immersive experience enhances understanding and facilitates on-the-spot decision-making.

Incorporating Solar Pumps

In the pursuit of self-sufficiency, homesteaders and rainwater enthusiasts are increasingly turning to solar energy as a sustainable power source. The integration of solar pumps into rainwater harvesting systems represents a significant leap toward energy efficiency, reducing reliance on traditional grid electricity while harnessing the power of the sun to optimize water circulation.

Solar-Powered Submersible Pumps

Submersible pumps play a pivotal role in transporting harvested rainwater from storage tanks to various points of use. Traditional electric pumps, while effective, contribute to energy costs and environmental impact. Solar-powered submersible pumps, on the other hand, utilize energy from the sun to drive water circulation. These pumps are not only cost-effective in the long run but also align with the ethos of eco-friendly, off-grid living.

Solar Panels and Battery Storage Systems

The heart of solar-powered rainwater harvesting lies in the integration of solar panels and battery storage systems. Solar panels convert sunlight into electrical energy, powering the submersible pumps directly or charging batteries for subsequent use during periods of low sunlight. Battery storage systems store excess energy generated during sunny days, ensuring a continuous and reliable power supply even during cloudy or nighttime conditions.

Sizing and Design Considerations for Solar-Powered Pumps

Optimizing the performance of solar pumps requires careful consideration of sizing and system design. Factors such as the daily water demand, pump capacity, sunlight exposure, and storage capacity all influence the sizing of solar panels and batteries. Properly designed solar pump systems maximize energy efficiency, ensuring a harmonious synergy between solar power and rainwater harvesting.

Solar Pump Efficiency Enhancements

Continuous advancements in solar pump technology contribute to enhanced efficiency and reliability. Innovations such as maximum power point tracking (MPPT) controllers optimize the power output of solar panels, ensuring the highest energy yield. Additionally, variable frequency drives (VFDs) enable precise control of pump speed, further optimizing energy consumption based on varying water demand.

Hybrid Systems for Redundancy

To address the intermittency of sunlight and potential energy gaps, enthusiasts may consider hybrid systems that integrate solar power with traditional electricity sources. These hybrid setups provide redundancy, ensuring a seamless transition to grid power or alternative energy sources when solar energy is insufficient. This approach enhances the reliability of the rainwater harvesting system, especially in regions with variable weather conditions.

Environmental and Economic Benefits

The integration of solar pumps into rainwater harvesting systems brings forth a myriad of environmental and economic benefits. By harnessing clean, renewable energy, enthusiasts reduce their carbon footprint, contributing to a more sustainable and eco-friendly lifestyle. Moreover, the long-term cost savings from reduced electricity bills and minimal maintenance make solar-powered rainwater harvesting an economically viable and environmentally responsible choice.

Landscape Design for Optimal Rainwater Use

The landscape that surrounds a homestead plays a crucial role in determining the efficiency of rainwater harvesting. Beyond the technical aspects of collection and filtration, thoughtful landscape design can maximize the utilization of harvested rainwater, creating a harmonious and sustainable ecosystem.

Permeable Surfaces for Water Infiltration

Traditional impermeable surfaces, such as concrete and asphalt, hinder rainwater absorption into the soil. In contrast, permeable surfaces, including permeable pavers, gravel, and grass pavers, allow rainwater to infiltrate the ground. This design choice promotes groundwater recharge, reduces runoff, and enhances the overall health of the surrounding ecosystem. Enthusiasts can strategically incorporate permeable surfaces in driveways, walkways, and patios to optimize rainwater absorption.

Rain Gardens for Natural Filtration

Rain gardens serve as natural filtration zones within the landscape, capturing and purifying rainwater before it enters the broader water collection system. These carefully designed gardens feature a combination of native plants, soil amendments, and mulch to enhance water absorption and filtration. By integrating rain gardens into the landscape, enthusiasts not only improve water quality but also create aesthetically pleasing and ecologically valuable outdoor spaces.

Contouring and Swales for Directed Water Flow

Incorporating contouring and swales into the landscape design directs rainwater flow in a purposeful manner. Contouring involves shaping the land to follow its natural slopes, preventing erosion, and promoting even water distribution. Swales, on the other hand, are shallow channels designed to collect and channel rainwater to specific areas, such as garden beds or storage tanks. These landscape features ensure efficient water utilization, preventing wastage and supporting the diverse needs of the homestead.

Green Roofs for Integrated Water Management

Green roofs adorned with vegetation offer a multifaceted approach to rainwater harvesting. Not only do they provide an additional surface for rainwater collection, but the vegetation also acts as a natural filter. The combination of green roofs with rainwater harvesting systems creates a holistic solution that enhances energy efficiency, reduces the urban heat island effect, and promotes biodiversity. This integrated approach to water management aligns with the principles of sustainable and eco-conscious living.

Multi-Functional Water Features

Innovative landscape design integrates water features that serve multiple functions. Decorative ponds, for example, not only enhance the visual appeal of the homestead but also act as additional rainwater storage. Combined with aquatic plants, these features contribute to natural filtration and support local wildlife. Enthusiasts can explore the integration of multi-functional water features to add beauty, functionality, and ecological value to their outdoor spaces.

Edible Landscaping for Dual Purpose

Aligning with the principles of permaculture, edible landscaping combines aesthetics with functionality. By strategically planting fruit and nut trees, berry bushes, and edible ground cover, enthusiasts create a landscape that not only produces food but also optimizes rainwater absorption. This dual-purpose approach transforms the homestead into a productive and resilient ecosystem, where every element serves a meaningful function in the broader context of sustainable living.

Community-Based Landscape Design Initiatives

Extending the benefits of landscape design beyond individual homesteads, community-based initiatives can transform entire neighborhoods. Collaborative efforts to implement permeable surfaces, rain gardens, and directed water flow systems contribute to a collective approach to rainwater harvesting. Community leaders can play a pivotal role in organizing and promoting such initiatives, fostering a sense of shared responsibility for sustainable water management.

CONCLUSION

As we reach the final pages of our journey together, pause and take a long, mindful look at the raindrops leaking from your gutters. Follow them cascading down the drainpipe into your freshly installed rain barrel. Do you view those droplets any differently now than before you began this book? Does the sight stir something within—a childlike wonder or a deeper connection to the natural cycles that sustain life?

When I first encountered rainwater harvesting years ago, I perceived it as merely a handy hobby—a small way to pinch pennies while gardening. But the more I researched, the more enchanted I became at the beauty of its premise—channeling the sky's endless bounty into a bottomless well of abundance. I saw it as the ultimate hack for a homesteader, conjuring independence, resilience, and self-reliance.

Hopefully, by now you share my enthusiasm, realizing just how profound an impact even basic rain harvesting makes. Every gallon collected reduces demand for tapped-out aquifers. Each system installed helps recharge local watersheds. And the combined efforts of enough homesteads could even balance regional water budgets. Like the iconic Butterfly Effect in chaos theory, a single rain barrel today cascades into a more sustainable tomorrow for all.

Yet the mission demands long-term commitment from each of us. I encourage you to revisit the maintenance chapter periodically to inspect your equipment. Expand and upgrade components whenever possible as your needs change. Consider investing in those advanced gadgets like self-cleaning filters or app-based monitoring once you master the basics. Treat your relationship with rainwater harvesting as a lifelong journey.

Remember also that sustainability requires a tapestry of earth-friendly practices woven together. Explore complementary systems like greywater usage, composting, solar energy, and regenerative agriculture. Discover the synergies between methods and implement what resonates. Lead through example in your community by sharing your expertise as a rainwater harvesting veteran.

While this concludes our main journey together, consider this a beginning rather than an end. I urge you to push further into the rabbit hole by seeking out additional resources beyond this book. Connect with fellow rainwater harvesting enthusiasts in forums and social networks. Experiment with new methodologies as innovative technologies emerge. Advance your knowledge through sites like the American Rainwater Catchment Systems Association, magazines like Home Power, or YouTube channels like The Urban Farmer.

Who knows, perhaps one day I will have the joy of leafing through a book that you author yourself, adding to humanity's accrued wisdom on harvesting rainwater. When that day arrives, I will celebrate the fact that a single raindrop of inspiration from my modest book sparked an entirely new wave of sustainable homesteading.

In closing, never lose sight of why we embarked on this journey to begin with—because water is life. It connects every organism across this wondrous, watery world. It flows through you and me, through soil and stone. Without it, all that we hold dear would cease. Rainwater harvesting gives each of us a chance to reciprocate the gift of water to become active participants in the never-ending water cycle.

I leave you with one parting thought. Next time you find yourself reveling beneath a soft spring rain, turn your face to the darkening sky with childlike wonder. Open your mouth wide as those sacred raindrops splash upon your tongue—tangy traces drawn up from the sea and flavored by the forest canopy. In that ephemeral moment, remember that what fell as rain may one day flow as milk within your goats, sweeten your strawberries, and nourish a towering elm shading your great-grandchildren. For all water is one water. And all rain is a promise waiting for visionaries like you to harvest.

Made in the USA
Monee, IL
30 July 2024